THE
THEORY OF EVERYTHING

The
Theory of Everything

A Nonperturbative,
Successfully Tested, Higgsless Theory
with
No Strings Attached

DONALD W. CALDWELL

DATUM
DB
BOOKS

To my wonderful wife, Jill

Table of Contents

Preface

Among the new concepts and theories in this monograph I shall introduce *The Theory of Everything*. That is, I shall derive and successfully test the unified field theory long sought by Albert Einstein that has become the Holy Grail of modern physics. In addition to this breakthrough, a more comprehensive overview of the scope and value of the book's content, as well as the motivation and purpose for writing it, can be found in the Introduction.

As an engineer and author unknown to the physics community, I expect many readers may be very skeptical about the book. In fact, they may be wondering if they would be wasting their time to read it. I have noticed that authors of physics books have occasionally been quite vocal, and shown disdain for outsiders who have been inspired to develop and publish new theories. Those outsiders have been referred to as "crackpots." Now, I too have read my share of today's strange and unusual theories, and while struggling through the numerous mathematical symbols and equations have quickly become disillusioned, and wondered if *I* were wasting *my* time. I have found, as others have, that many new, as well as long standing "mainstream" theories fail to convince me of any validity. However, before dismissing this book as a crackpot's inspiration, be assured the theories in this book are mathematically irrefutable and well worth your time to read them.

Although this book is primarily meant for professional physicists, students of physics, and those like myself who have a continued interest in the mysteries of physics from their college days, the results and conclusions may be understood by anyone curious about the nature of their universe. Understanding the new concepts, general discussions, and results and conclusions does not require complete comprehension of the mathematical derivations, which may be skipped by those without sufficient background. However, the only necessary requirements to follow the mathematics are a knowledge of the fundamental principles of mechanics and electricity, and an understanding of the methods of calculus including partial derivatives.

I have tried to minimize the number of equations where possible, yet include all significant steps of the derivations. However, the theories which led to functions too difficult or tedious to manipulate or calculate manually with pencil and paper were computed using Wolfram Research *Mathematica* software [1]. This software was also very convenient to plot the results.

During the course of developing the new theories, it was useful to quote examples of some passages from scientific texts which discuss existing theories and their conclusions in order to critically compare, contrast and comment on them, as well as provide references and a background for the discussion so the reader does not have to take my word for them, and can also easily understand the difference between the existing and new theories. It is not my intention to single out any of the individual authors of that material. Most of those examples are currently widely accepted as scientific fact.

I have put this work in the form of a monograph rather than publish individual journal papers to consolidate the theories and concepts in one place, and unify the presented postulates and hypotheses to help clarify their intended meaning. Unless acknowledged to be commonly found in existing textbooks, all of the derivations are new and original. I bear sole responsibility for its content.

On a more quantitative thought, due to my engineering background and preference, the rationalized mks system of units is used (almost) exclusively.

Frederick, Maryland
October 2019

D. W. Caldwell

Introduction

Unknown Causes

My interest in physics began in the early 1970s while I was an undergraduate student in mechanical engineering at the University of Maryland in College Park. Although I was not majoring in physics, the strange and mysterious phenomena and paradoxes in the physics courses to which I was exposed raised my curiosity. Particularly interesting were that the fundamental causes of many phenomena were completely unknown, and if there were attempts in textbooks to explain them, some appeared to me to be unsatisfactory. To name just a few, some of those strange mysteries and phenomena were:

> a. Gravity acted at a distance, and the cause was unknown.
> b. Objects were said to behave sometimes like particles and sometimes like waves.
> c. Electrons were said to go through two different slits at the same time.
> d. Photons were said to be massless particle-like wave packets which were emitted from atoms when electrons jumped to other orbits.
> e. The forces caused by electromagnetic fields could be calculated, but the cause was unknown because no one knew what an electric field was.
> f. Electrical charge was quantified as a property of some particles, but not physically defined.
> g. Some said particles gained mass as they increased speed, others said not.

The question of whether or not the mass of a physical object really increases with speed has been a controversy ever since the idea was introduced by Hendrik

Antoon Lorentz in an article in 1899. Some authors emphatically take one side or the other. Other authors find it useful to use each answer when convenient. The modern position of relativity theory seems to be that there is only one mass, the Newtonian rest mass, m_o, and it does not vary with velocity. It is said that when a mass at rest is accelerated to a relative velocity it doesn't gain mass, it just gains kinetic energy. Others believe mass increases with velocity, and call it relativistic mass.

Just as mysterious to me, as relativistic mass, was something commonly referred to as *charge*. The word charge was sometimes used to refer to an object, and other times used to refer to a property of an object. For example, charge was frequently associated with an electron, which was sometimes just called a charge. Also, an electron was said to contain, have, or carry a charge usually represented by the quantity -e, which no one could explain, as discussed by Kenneth Ford in his book, *The Quantum World*:

> We have no understanding of the particular magnitude of the charge quantum e. ... We don't know why it has the value it has. [2, p. 101]

Furthermore, the property charge also came in the "plus" variety, e.g. a similar type of particle called a positron. I shall derive an equation for the charge quantum e in this monograph which will precisely explain why it has the value it has.

The physical object charge was characterized in many ways. Not only could charge refer to the electron itself, but it was suggested that charge could be distributed throughout the electron's volume, or it may only cover its surface as if it were some strange substance that could reside at certain places. It was said that the electron itself could be a simple sphere of electric charge or a "charge cloud." Another possibility was that charge could be confined to a very small volume, perhaps at the center. However, if this were true, it was questioned: What holds together the strong mutual repulsion of its parts? In his book, *Relativity: The Special and General Theory*, Einstein thought what held it together may be gravitational forces:

> In the theoretical treatment of these electrons, we are faced with the difficulty that electrodynamic theory of itself is unable to give an

account of their nature. For since electrical masses of one sign repel each other, the negative electrical masses constituting the electron would necessarily be scattered under the influence of their mutual repulsions, unless there are forces of another kind operating between them, the nature of which has hitherto remained obscure to us.[1]

Notes

[1] The general theory of relativity renders it likely that the electrical masses of an electron are held together by gravitational forces. [3]

Some particles other than electrons and positrons were characterized as having charge, others were not. If a particle had it, forces were transmitted magically at a distance by something called a field, but charge wasn't concluded to be anything specific. It just *was*.

That, too, was the way of many physics phenomena. The effects were well known, named, and quantified, but no one knew what caused them. Richard Feynman states it this way in his book, *QED: The Strange Theory of Light and Matter*:

> ... while I am describing to you *how* Nature works, you won't understand *why* Nature works that way. But you see, nobody understands that. [4, p. 10]

The theories in this book shall allow you to understand "*why*."

In the intervening years since college, while reading sporadically to learn of new theories, and occasionally thinking about the mysteries as a hobby, I became aware of other even stranger hypothesized phenomena. However, I did not realize until recently, since retiring, as I read about the orthodox Copenhagen interpretation of quantum mechanics; reality not being real, but being pure mathematics, waves, and probabilities; that elementary particles are suggested to be strings; that one large particle, that lasts only briefly, is said to supply the mass to all other particles; and many other speculative theories without a shred of valid experimental evidence, just how convoluted the world of physics has really become.

Apparently, as a result of this state of physics, a number of authors have recently written books questioning the validity of many of those current speculative theories that are, however, claimed to be the best theories of today. I will not discuss those theories in any detail here. Many of them are generally well known from being popularized in the media. However, for those interested some of the theories and their issues are discussed in the following references [5][6][7][8][9].

Quantum mechanics is, in general, not one of the theories to which I refer above, but it will be impacted by the new theories in this book. So, I need to discuss a little bit about it and the effect it has had on physics.

Some Comments on Quantum Mechanics

On the positive side, quantum mechanics seems to be a well respected theory for making many correct predictions verified by experimental results. The main controversy surrounding it appears to be that its predictions are only the probabilities of finding particles, and that limitation has led to, let's say, *extreme* conclusions about our reality.

For example, quantum mechanics has led to the belief that Newton's laws of motion are not applicable to particles in microscopic systems. One such conclusion can be found in an introduction to quantum mechanics by Robert Eisberg and Robert Resnick in their well respected book, *Quantum Physics of Atoms, Molecules, Solids, Nuclei, and Particles*:

> We have presented experimental evidence which shows conclusively that the particles of microscopic systems move according to the laws of some form of wave motion, and not according to the Newtonian laws of motion obeyed by the particles of macroscopic systems. [10, p. 125]

It is this type of statement and others below from the field of quantum mechanics, which have created a disturbing mystical fog over physics, that I have difficulty accepting.

The main quantity with which quantum mechanics is concerned is a complex quantity called a particle's *wave function*. It is touted as the be-all and end-all to our knowledge of the microscopic world. Eugen Merzbacher phrases it like this in his book, *Quantum Mechanics*:

> Quantum mechanics contends that the wave function contains the maximum amount of information that nature allows us concerning the behavior of electrons, photons, protons, neutrons, quarks, and the like. Broadly, these tenets are referred to as the *Copenhagen interpretation* of quantum mechanics. [11, p. 18]

In this book this contention shall be proven to be false.

As I initially began reading books on quantum mechanics I couldn't help but notice that it is one of the more confusing and hyped areas of physics that I have encountered. It not only requires a significant background in many technical areas, and is itself difficult to understand, but also the originators and textbook authors themselves even state they don't understand it.

Although the wave function is given no physical interpretation, the square of the absolute value of its magnitude evaluated at a particular point and time is proportional to the probability of finding the particle there at that time. It is then said that when normalized, integrating the absolute value of the square of the wave function over all space must result in the value of one, since the particle must exist somewhere. It is also generally carefully stated that although the wave function is referred to as being spread out in space, this does not mean the particle itself is spread out. No part of the particle can exist at two different places at the same time.

This is all very reasonable, except the wave function is also described as a variable quantity characterizing the de Broglie wave which is said to "accompany" particles, and the de Broglie wave is also referred to as a traveling "matter" wave, where the matter is spread out in space. So, a particle, which cannot exist at two places at the same time, is said to be accompanied by a wave composed of matter spread out in space. This led to the concept of wave-particle duality where a particle may be partly described in terms not only of a particle, but also as a wave. Also, according to quantum mechanics, although a particle cannot exist in two different places at the same time, its wave function *can* exist in two different states at the same time. To make matters worse, the even more radical *Copenhagen interpretation* creeps in, which says the particle doesn't exist at all until measured, and that, in fact, an act of measurement creates the particle when the wave collapses! An example of this type of statement can be found in a definition of the Copenhagen interpretation, given in Manjit Kumar's excellent book, *Quantum*, which provides an extensive review of the subject:

> There is no quantum reality beyond what is revealed by an act of measurement or observation. Hence it is meaningless to say, for example, that an *electron* exists somewhere independent of an actual observation. Bohr and his supporters maintained that quantum mechanics was a complete theory, a claim challenged by Einstein. [12, p. 376]

Also, a couple of excerpts are presented below in which Kumar lucidly quotes a portion of Einstein's argument with Niels Bohr over his controversial conclusions about reality and the incompleteness of his theory:

> At the core of Einstein's physics was his unshakable belief in a reality that exists 'out there' independently of whether or not it is observed. 'Does the moon exist only when you look at it?' he asked Abraham Pais in an attempt to highlight the absurdity of thinking otherwise. The reality that Einstein envisaged had locality and was governed by causal laws that it was the job of the physicist to discover. [12, p. 352]

> ... Einstein wrote: 'I am, in fact, firmly convinced that the essentially statistical character of contemporary quantum theory is solely to be ascribed to the fact that this [theory] operates with an incomplete description of physical systems.' ... By 1954 he was adamant that 'it is not possible to get rid of the statistical character of the present quantum theory by merely adding something to the latter, without changing the fundamental concepts about the whole structure'. He was convinced that something more radical was required than a return to the concepts of classical

physics at the sub-quantum level. If quantum mechanics is incomplete, only a part of the whole truth, then there must be a complete theory waiting to be discovered. Einstein believed that this was the elusive unified field theory that he spent the last 25 years of his life searching for ... [12, p. 355]

Although not unanimously believed by all physicists today, the Copenhagen interpretation is said to be the generally accepted belief of the state of the microscopic world. Contemporary reading material contains many statements resulting from this interpretation, which are contradictory to a common "intuitive" picture of physical reality, that are now considered fact such as these below from various authors:

... for objects that are very small---electrons, atoms, photons---Newton's theory breaks down as well. With it, we also lose the concept of causality. The quantum universe does not possess the cause-and-effect structure we know from everyday life. [13, p. 13]

... quantum theory assumed that a particle *didn't have* a position until the measurement was made---it was the act of measurement that transformed its position from a probability to an actual value. And the same went for the other properties of the particle. [14, p. 21]

Nothing we have seen so far would enable us to calculate the velocity of a particle. It's not even clear what velocity means in quantum mechanics: If the particle doesn't have a determinate position (prior to measurement), neither does it have a well defined velocity. [15, p. 16]

Now Born and Bohr were saying that, in quantum physics at least, the reality had to be thrown away. *All* that existed was the probability. [14, p. 22]

... [To Werner Heisenberg] 'this objective world of time and space did not even exist'. [12, p. 356]

It's no wonder no one understands it. With all the outlandish conclusions that are drawn from the *Copenhagen interpretation*, it is difficult to separate what serious physicists actually believe about the subject, and what is just pure hype.

Like Einstein, some eminent contemporary physicists, while acknowledging that the statistical predictions of quantum mechanics agree well with experimental outcomes, question the completeness of the theory. In the last paragraph of his well respected textbook on the subject, *Introduction to Quantum Mechanics,* David J. Griffiths not only questions its completeness, but also its truth:

> In this book I have tried to tell a consistent and coherent story: ... But I cannot believe this is the end of the story; at the very least, we have much to learn about the nature of measurement and the mechanism of collapse. And it is entirely possible that future generations will look back, from the vantage point of a more sophisticated theory, and wonder how we could have been so gullible. [15, p. 433]

Also, while discussing how wave packets can be formed by superposing eigen-modes, and how these wave packets when quantized behave like particles, Anthony Zee questions the sufficiency of the "harmonic paradigm" used in quantum theory in his book, *Quantum Field Theory in a Nutshell*:

> It struck me as limiting that even after some 75 years, the whole subject of quantum field theory remains rooted in this harmonic paradigm, to use a dreadfully pretentious word. We have not been able to get away from the basic notions of oscillations and wave packets. Indeed, string theory, the heir to quantum field theory, is still firmly founded on this harmonic paradigm. Surely, a brilliant young physicist, perhaps a reader of this book, will take us beyond. [16, p. 5]

The insufficiency of the "harmonic paradigm" shall be demonstrated in this book.

Furthermore, the Nobel Laureate Gerard 't Hooft also rightly questions the restraints imposed on reality and the reasonableness of it in his book, *In search of the ulti-*

mate building blocks:

> In short, where is it [the electron] in reality? What is the reality that is
> hidden behind our formulae [quantum mechanics]? If we are to believe
> Bohr, it is senseless to search for such a reality. The quantum mechanical
> rules by themselves, and the actual observations performed by the detec-
> tors, are the only realities we are allowed to talk about.
>
> To this day, many researchers agree with Bohr's pragmatic attitude.
> The history books say that Bohr has proved Einstein wrong. But others,
> including myself, suspect that, in the long run, the Einsteinian view might
> return: that there is something missing in the Copenhagen interpretation. ...
>
> The elusive mystery of quantum mechanics gave rise to a great deal
> of controversy, and the amount of nonsense that has been claimed is so
> voluminous that a sober physicist does not even know where to start to
> refute it all. [6, p. 13]
>
> ...
>
> Much more reasonable is the suspicion that the statistical element in our
> predictions will eventually disappear completely as soon as we know
> exactly *the complete theory of all forces*, the Theory of Everything. [6, p.
> 14]

Purpose

This book is a scientific monograph of new results based on a combination of
generalized engineering practice and theoretical physics which provides "something
more radical" for a "start to refute it all," and "take us beyond." It will provide the
"ultimate building blocks," the reality hidden behind quantum mechanics, the
"something missing in the Copenhagen interpretation," and "the Theory of Everything."

In general, scientists believe the best physical theories have the greatest possible
generality, simplicity, and precision; characteristics lacking in many theories today.
However, the theories provided here will have all of those characteristics. None of
the derived equations shall be nonlinear. Perturbation theory shall not be required,

nor will strings be involved. No more than three spatial dimensions shall be necessary. I will not need to calculate any probabilities, or have a need for renormalization. Furthermore, I will not draw any Feynman diagrams, or need any of the mysteriously sounding unexplained force-related jargon commonly found in physics textbooks today such as: this particle "mediates" the force, "carries" the force, or exchanges the "virtual" particle. I will simply start over from the beginning, introduce two new basic postulates that are reasonable and easily understandable from *classical* physics, and *explain* many of physics' most mysterious phenomena.

Leave Your Preconceived Ideas Behind

To accomplish this goal I must simplify, yet severely alter, physics at the most fundamental level, and change the way the world is currently perceived. Consequently, I expect this may initially meet skepticism, and perhaps opposition, from some physicists deeply vested in other theories. As Max Planck is said to have stated:

> A new scientific truth does not triumph by convincing its opponents and making them see the light, but rather because its opponents eventually die and a new generation grows up that is familiar with it.

Therefore, I must request that while reading this book leave your current beliefs and personal preferences behind, be patient, and keep an open mind. By the end of Chapter 4, you will find that the time spent will be well rewarded by a new and amazing way to think of ourselves and the universe, validated by existing highly precise experimental data. It turns out that our world functions in such a way that some well known assumptions and theories, that have been thought to be true, will turn out to be wrong. Nothing is as it seems.

Furthermore, several other potential prejudices suggest that I request you to reserve judgment. As a result of the way physics has evolved over the past century in which today large amounts of funding are being provided to major ongoing experimental projects involving hundreds or even thousands of scientists to prove theories that have been deeply entrenched into the physics establishment, authors have stated that it is a "myth" that any single researcher could independently create any breakthroughs. I can only conclude from that negative thinking they themselves surely won't. Furthermore, I must point out that while physical experimental work is in many cases a "team sport," the original "thought" experiments of theoretical physics are generally not.

Also, although there may come a time when the ability to increase our knowledge of fundamental physics phenomena becomes asymptotically small, some authors seem to have taken the view now that nothing more can be discovered about some subatomic particles, and that we must be "content" with that. This conclusion may, in general, be based primarily upon the propaganda of the Copenhagen interpretation. However, I disagree, and shall show that conclusion is false.

Doubters Should Perhaps Jump to Chapter 1

In the next section there is a rather long list of some of the new and significant results which will be mathematically derived in this monograph. If I were to read this list when first beginning to read someone else's supposedly nonfictional scientific monograph, I may think this author *is* a crackpot, and toss the book aside. If you are likewise inclined, you may consider skipping the next section now and going directly to the first chapter. Returning later after you have read through the book will provide a better foundation for acceptance and belief. However, if you want a glimpse of what is ahead, continue reading, but be aware that you may strongly question the validity of what you read.

Scope of New and Significant Results

In this book you shall find the following new and significant results and conclusions:

(1) An evidence-based unified field Theory of Everything shall be derived based firmly on a new mechanical potential function and electrodynamic theory.

(2) It shall be concluded that all matter is constructed from only one *single elementary* particle. It will not be exotic or contain any mysterious properties like "strangeness" or "isospin." It will be shown to be a very simple building block of all other particles, and the cause of all the forces between all particles. The 60, or so, particles that are said to exist as "elementary" particles in the standard model are determined not to be elementary, but be composite particles. This theory shall cause a fundamental revolution in elementary particle physics, and confirm that Paul Dirac was correct, while referring to the electron and positron, when he said that two fundamental particles are one too many.

(3) A new theory of force shall be derived, and eliminate the hypothesized mysterious phenomenon of two-way transmission of particles between objects, commonly referred to as "particle exchange." It shall be explained why oppositely charged particles attract and like charged particles repel, matching Coulomb's Law. This force theory will turn out to be an example of theoretical reduction, the process by which one theory reduces another theory to more basic terms. It will reduce the four so-called fundamental forces into one fundamental force at the heart of all the others, unifying all of the four so-called separate "fundamental" forces.

(4) A new theory is derived to explain why particles have a macroscopic mass. It will be called Restraint Density Theory (RDT).

(5) A fundamental theory of the electron and positron shall show that although they sometimes behave like particles and sometimes waves, they are neither. Until this book, the mass of the electron, the smallest non-zero rest mass of all known charged particles, was not calculable from any basic fundamental principles. Its value was attained by measurement. Furthermore, the electron was thought to be structureless, and the reason for it having a definite radius unknown. However, in this book I shall derive from *fundamental principles* the structure and mass of the electron (and positron), which matches precisely its experimentally determined characteristics and value, and explain why it has a definite radius. This has never been done before for any particle. Furthermore, the electron's mass shall be shown to be entirely electromagnetic, i.e. entirely due to its electromagnetic field.

(6) The law of conservation of mass and energy shall be shown to *not* be

required in our three-dimensional universe, i.e. specifically, it does not apply to electrons and positrons.

(7) The quantity called "charge" shall be shown *not* to be a strange mystical substance, but simply a *measure of mass flow rate.*

(8) The precise nature of electric fields shall be determined by deriving a quantitative relationship between electric field strength and mass flow rate; two phenomena formerly believed disparate, and a relationship emphatically denied.

(9) The photon shall be shown to *not* be the fundamental elementary particle of electric fields, but simply be a disturbance in an electric field.

(10) Electric fields are determined to be made of quantum particles of mass, which are *not* subject to the force of gravity.

(11) The movement of macroscopic objects will be shown to be caused by the collision of fields of quantum particles of mass.

(12) A new form of Bohr's theoretical Rydberg equation is derived which provides the key to determining the cause of the frequencies of photons.

(13) A new model of the atom, an extension to the Bohr theory, is derived.

(14) The missing piece of quantum mechanics is discovered. The probabilis tic foundation of quantum mechanics will be superseded by a deeper level of physical reality.

(15) It is demonstrated that causality and an objective world of time and space are alive and well. In short, the Copenhagen interpretation is generally nonsense.

(16) The composition and structure of photons, and the photon-creation process shall be explained.

(17) Bohr's *correspondence principle* shall become obsolete.

(18) The measurement problem in quantum mechanics, the question of how (or whether) a wave function collapse occurs, shall be answered.

(19) It is concluded that a particle's wave function cannot be in two different states at the same time.

(20) It is shown that matter is the result of a fluid flow process, in opposition to Louis de Broglie's hypothesis that matter is to be regarded as a wave process.

(21) It shall be concluded that so-called massless particles which have momentum do not exist.

(22) The cause for the limiting speed of light shall be determined.

(23) The phenomena of electron-positron annihilation and pair production shall be mathematically derived. At small distances two oppositely charged point charges merge together and form a new combined particle before annihilating. This merging process shall be sequentially diagrammed.

(24) Rest volume, rest mass, and rest density of electrons and positrons shall be shown to be variable quantities, not constants.

(25) The phenomenon known as the mass defect inside nuclei shall be hypothesized to be due to the change in the mass density field formed from the combination of their constituent particles.

(26) It is concluded that the electrostatic forces on point charges diverge significantly from Coulomb's empirical force law at small separation distances.

(27) It is shown that the magnitudes of the repulsive and attractive electrostatic forces between two like charged and two oppositely charged point charges, respectively, at the same separation distance are generally *not* equal.

(28) An answer to the mystery of the universe's missing anti-matter shall be hypothesized.

(29) Definitions of matter and radiation, not previously well defined, shall be proposed.

(30) A new theory called Relativistic Restraint Density Theory shall provide the *first* fluid dynamics justification for the original hypothesis by H. A. Lorentz, that the form of the electron experiences a contraction in the direction of motion.

(31) Relativistic Restraint Density Theory shall validate the concept of relativistic mass, and suggest the end of the argument between physicists as to whether the mass of macroscopic objects really increases with speed.

(32) The universal validity of the relativistic mass equation, $m = \gamma m_0$, shall be questioned, and it will be concluded that *all* relativistic equations derived from it may be no more than approximations.

(33) A new relativistic kinetic energy equation, based on the derived unified field theory, fits existing data significantly better than Einstein's accepted theory based on "ponderable masses."

(34) Gravity shall be determined to be an electromagnetic phenomenon.

(35) It is concluded that the force of gravity does not attract light.

(36) The cause of Dark Energy shall be hypothesized.

(37) If "Dark Matter" exists, it is just ordinary matter.

(38) The validity of the Big Bang theory shall be questioned.

... among other major results and conclusions.

Chapter 1

Starting Over from the Beginning

This monograph contains important fundamental breakthroughs that will change the course of science. It introduces several key concepts that lead to precise quantitative results, and provides answers to many of the most difficult questions in all of physics, as well as dispelling many of the shocking conclusions reached over the last century. As a result, many current beliefs, teachings, and highly publicized theories shall be rendered unsupportable.

To accomplish this, it will be necessary to consider some basic thoughts about what causes forces. To begin, I shall be mainly interested in determining the precise fundamental description of electrons and positrons, their nature, and how they interact and exert forces on each other. To do this it will be necessary to introduce some new basic assumptions and concepts. However, before doing that let's initially consider how forces are currently thought to be transmitted between two objects.

Particle Exchange and Exchange Force

Many books state that the force between two separated bodies is caused by a (seemingly unexplainable) phenomenon called "particle exchange." This is a hypothesized phenomenon where objects or particles called "force carriers" are said to be sent back and forth from one body to another to cause attractive and repulsive forces between them. Each of the four conventionally accepted fundamental forces: electromagnetic, weak nuclear, strong nuclear, and gravitational are said to have their own unique "force carrier" particle to exchange.

This phenomenon doesn't seem to be well understood, because no one seems to be able to provide a reasonable explanation of it. The typical statements in the several references below which discuss it don't help very much to clarify it. However, for now let's just look at what some authors say about it. Please note, these descriptions are not unique to these authors, but are generally accepted.

As Arthur Beiser states in his textbook, *Perspectives of Modern Physics*:

> ... there is no simple mathematical way of demonstrating how the exchange of particles between two bodies can lead to attractive and repulsive forces ... [17, p. 512]

After this statement, he then shows a figure containing two basketball players facing each other with their arms outstretched exchanging basketballs, and claims that both attractive and repulsive forces can arise from the exchange of these basketballs, i.e. particle exchange. If a basketball is thrown from one player to another, the repulsive force between players is certainly easy to understand and accept, but he also states that attraction can result in the following manner:

> If the boys snatch the basketballs from each other's hands, however, the result will be equivalent to an attractive force acting between them. [17, p. 513]

I don't mean to be rude, but I think it is safe to say that microscopic "bodies" don't have arms and hands to exchange particles or transfer momentum.

However, he then asks the excellent question about so-called "force carriers" between nucleons, called mesons, that supposedly transmit a nuclear force:

> If nucleons constantly emit and absorb mesons, why are neutrons or protons never found with other than their usual masses? [17, p. 513]

He attempts to explain this by invoking experimental limitations, and stating that a single nucleon receives and transmits mesons at *nearly* the same time:

The laws of physics refer exclusively to experimentally measurable quantities, and the uncertainty principle limits the accuracy with which certain combinations of measurements can be made. The emission of a meson by a nucleon which does not change mass ... can occur provided that the nucleon absorbs a meson emitted by the neighboring nucleon it is interacting with so soon afterward that *even in principle* it is impossible to determine whether or not any mass change actually has been involved. [17, p. 513]

Note that this description of particle exchange has both of the interacting nucleons emitting *and* absorbing "force carriers."

There are many authors who discuss the throwing of basketballs to cause repulsion. It has been more difficult to find references which discuss a better understanding of how attraction occurs. In the preface to the Dover Edition of the book, *Concepts of Force*, the author, Max Jammer, after discussing an exchange of balls by skaters on a frozen lake to produce the repulsive force, says that:

> ... an exchange of boomerangs, hitting each of them from the outside (or the exchange of negative momenta), will produce an attractive interaction. [18]

The boomerang concept does seem to make some physical sense, but if this is to apply to electrons and positrons, then each must have the ability to throw *two* different types of "force carriers" which change depending upon which type of object they confront to create both attractive and repulsive forces.

It has been said that "virtual" photons are the "force carrier" particles exchanged in the case of attraction between an electron and positron. Robert D. Klauber in his textbook, *Student Friendly Quantum Field Theory,* asserts that in the case of attraction, a:

> ... virtual photon carries 3-momentum in the opposite direction of its velocity. This certainly seems strange from a classical mechanics perspec-

tive, but it is the only way for attraction to occur. [19, p. 244]

I agree that this explanation "certainly seems strange." It is very disturbing to me that an assertion is made that a particle with such a fundamental directed property as linear momentum can transmit it to another particle in the *opposite* direction of its own velocity. This completely contradicts the conservation of momentum, a most sacred principle in physics. Amazingly, this assertion seems to be the general consensus among physicists, part of the so-called *standard model*, and taught today.

However, in his book, *Introduction to Quantum Mechanics,* David J. Griffiths discusses the exchange force as not really a force at all, but a consequence of the wave functions of two particles:

> The *interesting* case is when there *is* some overlap of the wave functions. The system behaves as though there were a "force of attraction" between identical bosons, pulling them closer together, and a "force of repulsion" between identical fermions, pushing them apart ... . We call it an **exchange force**, although it's not really a force at all --- no physical agency is pushing on the particles ... . It is also a strictly quantum mechanical phenomenon, with no classical counterpart. [15, p.209]

The last sentence in this quote frequently occurs in books on quantum mechanics. It seems to be used to avoid any precise explanation of a phenomenon currently unexplainable. The phenomenon is said to be "a strictly quantum mechanical phenomenon, with no classical counterpart."

In the paper *Exchange Forces*, the author, David Falkoff, states that:

> More precisely, the interaction between two charges can be interpreted as due to the emission of photons by the one charge and subsequent absorption of these photons by the other. ...

> The above description must be qualified in one important respect. As it stands, it seems to predict, incorrectly, that one should observe

radiation from charges at rest. To remove this objection, it is necessary to draw (the purely quantum-mechanical) distinction between real and virtual photon emission: Namely, a virtual photon emission (or absorption) can be characterized as one which cannot be detected because it would violate the energy conservation law. For example, an electron at rest cannot emit a photon of any frequency v, having energy hv and momentum hv/c, because it would be impossible for the electron to remain at rest and still maintain the constancy of energy and momentum for the system of electron plus photon. But, it is convenient, in this case, to speak of virtual photon emission. [20, p. 35]

Notice here, one charge emits and the other charge absorbs, and the emitting charge must emit two different types of photons which depend on whether a charge is at rest or in motion. Furthermore, there is a requirement that the "virtual" photon emission or absorption cannot be detected.

Each of the authors above have described different, and rather unsatisfying causes of the forces between bodies. This appears to be the state of understanding this mystery today. In order to solve this problem, I will later mathematically derive a new, but somewhat similar, concept that will explain distinctly and exactly the causes of both repulsive and attractive forces between bodies. For now, let's clear our minds, go back to basics, and briefly consider some important experimental results and initial assumptions necessary as a premise for new theories.

Experimental Results and Initial Assumptions

There is a well known experiment that needs to be emphasized as a basis to justify the derivations which follow. It is the Michelson-Morley experiment conducted in 1887. It attempted, but failed to find a unique inertial "ether" system. Instead, the experiment helped to establish that the speed of light in free space is a true constant of nature. Here I will rely on the results of the Michelson-Morley experiment, from which it is concluded there is

(1) no unique fundamental inertial "ether" system to transmit wave momentum,

and

(2) the speed of light in free space, c, is a true constant of nature.

In addition to the above two *experimental* conclusions, I will assume the validity of the principles of causality and locality. That is, I will assume the requirement that a cause precedes its effects, and that an event that happens in one place shall not cause whatever happens at a distant location that cannot be reached in time by traveling at the speed of light in free space. That is, there is no instantaneous transmission of action at a distance.

The principle of locality is occasionally supposedly "discredited" when considering a suggested phenomenon called quantum entanglement. Conclusions from the hypothesis of quantum entanglement state that *instantaneous* changes can occur to one particle caused by another particle when they are separated by large distances. Apparently, some physicists believe experiments support this. However, in elementary particle physics all interactions are assumed to be local. Since I will be using the principle of locality to support my derivations, I will make the assumption that the so-called phenomenon of instantaneous actions or "spooky actions" at a distance is unsupportable, and rely solely on *particle collisions* to transmit forces.

The Phenomenon of Collisions

When Sir Isaac Newton developed his laws of motion, he frequently used the translated words *exerted* and *impressed*. However, he did not specifically say how forces of attraction or repulsion are transmitted, but rather avoided those uncertainties, and confined his work to the results of such transmissions. The following excerpt discussing some possible causes of force transmission comes from Florian Cajori's translation, *Newton's Principia*:

I here use the word *attraction* in general for any endeavor whatever, made by bodies to approach each other, whether that endeavor arise from the action of the bodies themselves, as tending to each other or agitating each other by spirits emitted; or whether it arises from the action of the ether or of the air, or of any medium whatever, whether corporeal or incorporeal, in any manner impelling bodies placed therein towards each other. In the same general sense I use the word *impulse*, not defining in this treatise the species or physical qualities of forces, but investigating the quantities and mathematical proportions of them; ... [21, p. 192]

From this historical perspective I will begin by proposing a specific cause for the interaction that exerts *all* known forces, and call it the Collision Principle.

Collision Principle:

The only way a force can be directly transmitted from one body to another is for the two bodies to physically collide, i.e. contact.

Of course, forces need not be transmitted *directly*, but can be *indirectly* transmitted. That is, one object can impact another object which in turn through a series of one or more impacts transmit forces, but fundamentally a force cannot be *directly* transmitted without two objects physically colliding or contacting in some manner. Thus, all persistent forces operating within any system must be transmitted through the contiguous actions of collisions. This may seem obvious for repulsive phenomena, but may certainly seem less likely for attraction.

When Newton introduced his law of gravity which acts at a distance without any apparent explanation, it certainly would have raised some questions at that time, as it may now, if his other suggested potential causes had been ruled out, and collisions were specified to transmit an attractive force. However, here I will arbitrarily rule out all other potential causes without proof, and assume the common notion of the principle of collision to transmit all forces, both repulsive *and* attractive. This will further reinforce the principle of locality, and more definitize how fundamental forces are transmitted.

What is a Point Charge and What is Its Field?

Since I shall be specifically interested in examining how the Coulomb force between two separated point charges is transmitted, and since the two point charges *themselves* obviously do not collide for this force to occur, there must be something more happening between them to cause this interaction. In general, this "something" has been called an electric field. But, fundamentally what is a point charge and what is an electric field?

The two concepts "electric charge" and "electric field" are extensively used, and generally taken for granted. Their effects have been well quantified, but they have never been adequately explained because no one has known enough about their fundamental nature to conclude what they are. This is explicitly stated in the following excerpts taken from Robert Oerter's *The Theory of Almost Everything*, K. Ford's *The Quantum World*, and D. Griffiths' *Introduction to Electrodynamics*, respectively:

> Our inability to define the charge in any absolute sense is the ignorance that gives rise to a quantum field: the photon field. [22, p. 185]

> Electric charge is that certain something (that *je ne sais quoi*) which makes a particle attractive to the opposite kind of particle. ... why don't the bits of charge that make up its total charge cause it to fly apart? Are space and time drastically warped right at the particle's location? These may not even be the right questions to be asking, but they suggest why we can't say that we fully and deeply understand charge. [2, p. 22-23]

> What exactly *is* an electric field? ... I can't tell you, then, what a field *is*--- only how to calculate it and what it can do for you once you've got it. [23, p. 61-62]

So, I shall assume that at this point no one knows the true nature of either an electric charge or an electric field.

Now, it is universally accepted that electrons and positrons have electric charges and electric fields. In fact, they are commonly referred to as *point charges*. In some texts the words *point charge* may also be used to refer to a small single object that is said to be "charged," or a collection of charges in a very small volume. I will not use it in either of those ways. To avoid any confusion I will use the phrase *point charge* only to refer to electrons and positrons. Furthermore, this is distinguished from a *point particle* which is defined as an extremely small particle of infinitesimal size with mass that does *not* have an associated charge or electric field.

With this fundamental background to help deduce explanations of what the electric charge and electric field of an electron and positron actually are, I shall next need to briefly examine the somewhat, but not exact, analogous singularities of sources and sinks that are used in the engineering field of incompressible fluid dynamics. Since I assume most readers have had at least one elementary course in this area, I have tried to minimize the background discussions to only those necessary for clarity.

Chapter 2

A Fluid Analogy

Fluid Singularities

One of the major concerns in the mathematics used in engineering and physics has been the existence of singular points. These points are generally not allowed within practical solutions to problems because of the infinities theoretically predicted there.

Point charges have been shown *experimentally* to exhibit properties like singularities. That is, the magnitude of their electric fields tend toward infinity, and theoretically become infinite at their centers. Given the spherical coordinate system (R, θ, ϕ) in Figure 2-1, the electric field vector, **E**, of a single positron situated at the origin is given by the vector equation

$$\mathbf{E}(R) = \frac{ke}{R^2}\,\hat{\mathbf{R}} \tag{2-1}$$

where e is the elementary charge, k is the Coulomb force constant, R is the distance from the origin, and $\hat{\mathbf{R}}$ is the unit vector in the radial direction. The electric field of an electron is simply the negative of that of the positron.

In the field of fluid dynamics two useful types of singularities are called sources and sinks. A source is a fluid flow field where the fluid velocity vectors are directed away from a central point, and a sink is a flow field where the velocity vectors are directed toward a central point. The center of a source may be considered as a point from which fluid issues at a constant mass flow rate and flows along radial paths, and the center of a sink may be considered as a point toward which fluid flows at a constant mass flow rate along radial paths and disappears.

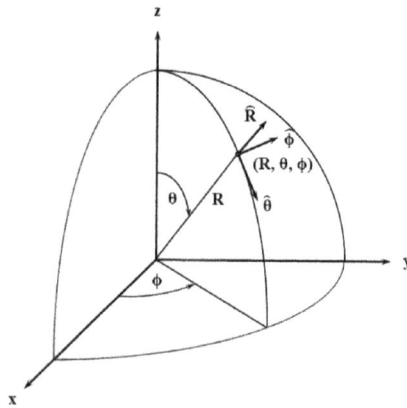

Figure 2-1. A spherical coordinate system

Sources and sinks have been constructed to be either two-dimensional or three-dimensional flows to solve problems consisting of specifically *incompressible* fluids. These fluid singularities are useful mathematical concepts because *they cause forces* which are used to change the flow patterns of fluid fields. From the conservation of mass in the fields, the three-dimensional flow equations have an inverse R-squared dependence similar to that of point charges. For these flow patterns the magnitude of velocity tends to infinity toward their centers, and theoretically becomes infinite there. The velocity field vector, **V**, of a single source situated at the origin of the spherical coordinate system is given by the vector equation

$$\mathbf{V}(R) = \frac{Q}{4\pi R^2}\,\hat{\mathbf{R}} \tag{2-2}$$

where Q is the constant volume flow rate. The velocity field of a sink is simply the negative of that of the source. It is this analogy between fluid sources and sinks, and positrons and electrons that has led me to a hypothesis about their relationship in the next section.

"Something More Radical"

Now, since the Coulomb force between separated point charges is *not* caused by their *direct* collisions, from the *Collision Principle* the electromagnetic field must not be empty, but must really be a space highly populated in some manner with objects of *mass* streaming between the separated point charges that either collide directly, or in combination with other objects to transmit the electrostatic force.

Since *incompressible* fluid singularities, which constantly emit or absorb fluid particles, have been used in classical fluid dynamics texts to cause forces that change the motion of fluid flow patterns, and they are at least somewhat, but not exactly, analogous to point charges, I shall propose my first hypothesis to connect their mass fluid flow phenomenon to point charges in order to serve as a motivation for the analysis to follow.

Hypothesis I:

Point charges, i.e. electrons and positrons, constantly either emit or absorb objects, that <u>*have mass*</u>, *along radial paths in all directions. It is unknown at this time whether the objects always stream away from or toward positrons, but whichever applies the opposite is true for electrons.*

As discussed earlier, currently accepted theory states that forces are transmitted by particle exchange. For point charges the so-called "force carriers" are said to be photons, where it is said that "the quantized electromagnetic field is made up of a collection of identical particles called photons" [24, p. 475]. As made clear in the hypothesis, I am *not* suggesting the objects emitted or absorbed by electrons or positrons are any type of real or so-called "virtual" photons. On the contrary, it will become evident as I proceed that I'm suggesting that electromagnetic fields are composed of a much smaller *sub-photonic* particle, that *has mass*, from which photons are created when electrons change orbits. This particle is theorized to be the *true* quantum of the electromagnetic field, *not* the photon. But, I'm getting ahead of myself. So, let's continue by considering an implication of the first hypothesis.

In the hypothesis, I am explicitly violating the law of local conservation of mass and energy, where it is said that in any closed system mass and energy can neither be created nor destroyed. In a closed system composed of a control volume enclosing a source, both mass and kinetic energy are created, and in a closed system composed of a control volume enclosing a sink, both mass and kinetic energy are destroyed. These two examples clearly violate the principle of local conservation of mass and energy, at least in the three-dimensional world in which we appear to live. Therefore, to accept this hypothesis we must reject the assumption of *local* conservation of mass and energy.

This may be difficult to accept, but the rejection of long accepted assumptions is frequently necessary in developing new theories. Lee Smolin persuasively phrased this while criticizing some current theories in his book, *The Trouble with Physics*:

> ... all theories that triumphed had consequences for experiment that were simple to work out and could be tested within a few years. ...
>
> Whatever else one says about string theory, loop quantum gravity, and other approaches, they have not delivered on that front. I believe there is something basic we are all missing, some wrong assumption we are all making. If this is so, then we need to isolate the wrong assumption and replace it with a new idea. [5, p. 256]

A. Zee, in his closing words while discussing potential new research in his book, *Quantum Field Theory in a Nutshell*, clearly foreshadowed the same need:

> In all previous revolutions in physics, a formerly cherished concept has to be jettisoned. If we are poised before another conceptual shift, something else might have to go. Lorentz invariance, perhaps? More likely, we may have to abandon strict locality. ... But we need analyticity, and of course analyticity follows from locality and causality, as far as we understand. [16, p. 521]

Thus, if Hypothesis I is proven to be true, then one wrong assumption has been the law of *local* conservation of mass and energy in our three-dimensional universe.

That is, particles with mass and energy may be continuously created or otherwise produced at the center of a source-like point charge, and continuously annihilated or otherwise disappear at the center of a sink-like point charge.

If there were a *constant* equal number of sources and sinks of equal flow rate in the universe, this would support a global conservation of mass and energy, but not local. That is, particles with mass and energy may be continuously created or otherwise produced at the center of a source-like point charge, and continuously annihilated or otherwise disappear at the center of a sink-like point charge to create a *one-way* fluid flow between them.

Therefore, a consequence of Hypothesis I is:

Corollary to Hypothesis I:

The law of conservation of mass and energy in a closed system is not required in our three-dimensional universe. In fact, it is violated by point charges.

However, some physicists contend that even though a phenomenon may actually occur that would violate the conservation of energy law, if it cannot be detected the law is not violated. This was illustrated in the statement by David Falkoff quoted in Chapter 1, "a virtual photon emission (or absorption) can be characterized as one which cannot be detected because it would violate the energy conservation law." I, rather, prefer to hold accountable even undetectable phenomena that actually occur to our universal laws.

Sign Convention and Other Matters

At this point, I need to say a little bit about sign convention to avoid potential confusion and help clarify things to come. In the mid-eighteenth century, Benjamin Franklin first introduced positive and negative designations for the two types of electrical charge. Which charge was chosen to be called positive and which negative was entirely arbitrary. The electron's charge was given the negative sign. Later, the positron was found, and assigned the plus sign. Subsequently, the positively charged positron has been described as *antimatter*, and the negatively charged electron has been described as *matter*.

In this book I will be introducing three-dimensional sources and sinks which have the positive and negative connotation of assigned flow direction, respectively. Because of the assigned sign convention, the positron should be paired with the source, and the electron should be paired with the sink. However, there may be some confusion when it is realized that a source that produces mass is called antimatter, and a sink that eliminates mass is called matter.

In order to be clear throughout, I will stay with the historical conventions that align electrons and sinks with the negative sign, and positrons and sources with the positive sign. It should be kept in mind, however, that later when it is shown that in reality positrons and electrons behave precisely as a special type of source and sink, it is really unknown which one is producing mass and which one is eliminating mass because of the arbitrary nature of the original sign assignment. Wherever possible, I will use the term "point charge" where I can refer to both the electron and positron without having to designate a sign.

Also, historically, matter has not been well defined, and does not have a universally accepted definition. It is commonly stated that the universe is composed entirely of matter and radiation. In this sense, anything that is not radiation is matter, and antimatter is not specifically distinguished from matter. Because of this, when I refer to matter I will generally mean objects which are not simply radiation.

With these bits of potential confusion, and hopefully clarification behind us, let us continue with a small section on basic introductory fluid dynamics before we begin with the first necessary derivations of fluid flow fields.

Basic Fluid Dynamics Background

On a macroscopic scale, a fluid may be thought of as a continuum. However, the concept of a continuous medium as applied to a fluid is an idealization. A fluid on the microscopic scale is composed of discrete particles or molecules. Because of these two views of fluids, there are two methods which are used to describe fluid motion: the Lagrangian method and the Eulerian method.   In the Lagrangian method, we follow the motion and paths of each fluid particle of fixed identity. This is similar to the tracking of particle trajectories which is performed at large particle accelerator facilities.   In the Eulerian method, we specify continuum properties of the fluid as functions of position and time, and simply describe what is happening at different positions at given times.   In dealing with engineering problems on the macroscopic level, the volumes under consideration are very large and contain many molecules.   Therefore, the Eulerian method is mathematically more convenient in the study of fluid dynamics, and it will also be used exclusively when deriving the fluidic equations in this monograph.

In the next chapter, we will be examining incompressible fluid flows of a source field, sink field, and both of them together in a combined field. These flow fields will be simple ideal irrotational flows known as *potential* flows which are characterized by functions called velocity potentials.   Once we derive the velocity potential for the combined source-sink field, we will determine the force between the source and the sink as a function of separation distance. The fluid force will then be compared to the Coulomb force between a positron and electron at the same separation distances. This will all be accomplished and explained in some detail.

Chapter 3

The Force Between An Incompressible Source and Sink

Although *incompressible* fluid singularities are useful mathematical concepts which cause fluid forces that are used to change fluid flow patterns, they have no physical counterparts since the magnitudes of velocities at their centers are theoretically infinite. The motion of a flow region caused by such singularities would have a physical meaning only when the region enclosing them can be excluded from consideration. In engineering practice, this is accomplished by creating a new flow pattern by putting sources and sinks in a flow field to create a representation of a useful rigid body around them. The main problem is to find a streamline pattern which will serve as the boundary of the body. Then, the boundary and flow field external to the body are used to, for example, derive the stream function and velocity potential for the flow, and determine stagnation points, pressures, and perhaps hydrodynamic drag on the body.

However, for the problem at hand I will *not* create a flow pattern to represent a rigid body. Instead, I will develop a simple fluid flow field containing only one incompressible three-dimensional source and sink. I will then determine the field's velocity potential, create a spherical control volume around the source, and solve directly for the force on the fluid within the control volume.

Although I believe this specific problem is new and interesting in its own right, the reason for developing this *incompressible* derivation and method of solution now is to generalize it later in Chapter 4. There, I shall introduce the new concepts of a three-dimensional *compressible* fluid source and sink, and derive and solve the equations of the fluid flow field containing them. The compressible source and sink will be shown not only to be useful mathematical concepts, but also have physical counterparts as positrons and electrons.

If the reader is very familiar with basic incompressible fluid methodology, he or she may have to exercise a little patience while I discuss some of it in the next few sections, and then attain the results and conclusions for the problem. However, developing the *incompressible* fluid methodology first, based on already established and commonly used principles, will make the new generalization to *compressible* flow much easier to follow and understand in Chapter 4.

Differential Equation of Continuity

Historically, during the general process of developing the fundamental equations of fluid dynamics, and *before* restricting fluid flows to be incompressible, the following differential equation of continuity is derived which is valid for *any* fluid flow [25, p.64, (2-15)]:

$$\mathbf{\nabla} \cdot \rho\mathbf{V} + \frac{\partial \rho}{\partial t} = 0 \qquad (3\text{-}1)$$

where $\mathbf{\nabla}$ is the vector gradient operator, ρ is the mass density function, $\mathbf{V}$ is the velocity vector, and t is time. Any mathematical function which represents a possible case of fluid flow must satisfy this equation everywhere in the flow field except at points of discontinuity where the derivatives are not defined.

Fortunately, we will not need to be interested in flows which vary with time, and for the case of steady flow, the equation simplifies to

$$\mathbf{\nabla} \cdot \rho\mathbf{V} = 0 \qquad (3\text{-}2)$$

Idealizing the Fluid as Incompressible

At this point in the development of elementary fluid dynamics, the simplest and common path forward is to idealize the fluid as incompressible. Under the assumption of incompressibility the fluid field equations simplify greatly, facilitating the solutions to many practical problems. To do this the density is specified as a constant, and it then drops out of the continuity equation, which becomes

$$\nabla \cdot \mathbf{V} = 0 \qquad\qquad (3\text{-}3)$$

Also, important simplifications which result in the ability to analyze many fluid flow patterns occur by introducing the concept of irrotational flow. In an irrotational fluid flow all of the fluid elements do not rotate during their motion. The condition for a fluid flow to be irrotational is that the **curl** of the velocity vector, $\mathbf{V}$, equals zero:

$$\nabla \times \mathbf{V} = \mathbf{0} \qquad\qquad (3\text{-}4)$$

Aside: *Interestingly and importantly*, static electric fields are also irrotational fields. In any static electric field the **curl** of the electric field vector, $\nabla \times \mathbf{E}$, equals zero for *any* static charge distribution whatsoever.

Assuming an irrotational fluid flow, a point function called the *velocity potential*, Φ, can be defined where

$$\mathbf{V} = \nabla \Phi \qquad\qquad (3\text{-}5)$$

That is, the velocity field is always expressible as the gradient of some scalar function Φ.

Substituting equation (3-5) into (3-3) results in

$$\nabla \cdot \nabla \Phi = 0 \qquad\qquad (3\text{-}6)$$

or,

$$\nabla^2 \Phi = 0 \qquad\qquad (3\text{-}7)$$

This type of equation is called a Laplace equation [25, p.145, (3-34)].

In the derivation in this section we have required the fluid flow to be steady, incompressible, and irrotational. This has led us to the requirement that flows of this type must satisfy the Laplace equation. Since the Laplace equation is linear, solutions can be easily attained, and the superposition of those solutions is possible to solve other flow problems. In the next section we will develop the velocity potentials for an incompressible source and sink which satisfy the Laplace equation.

A Three-Dimensional Incompressible Source and Sink

Both two-dimensional and three-dimensional source and sink flow patterns are commonly used in fluid dynamics problems associated with ideal incompressible flow. In this book, we will only be using those with three-dimensions. In this case, the sources and sinks are characterized as having a constant volume flow rate, Q, issuing from their centers and toward their centers, respectively. Consider a three-dimensional incompressible fluid flow of a source in a fixed inertial reference frame with streamlines radiating from a central point located at the origin of a reference z-axis as shown in Figure 3-1.

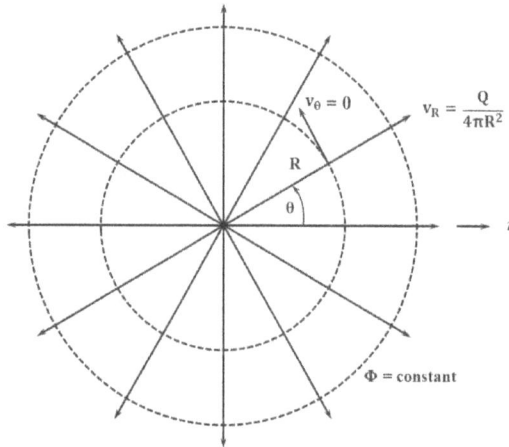

Figure 3-1. A three-dimensional incompressible source.

We will do our analyses in the spherical coordinate system (R, θ, ϕ) with the angle θ opening relative to the reference z-axis. Fortunately, we will only need to examine relatively simple three-dimensional fluid problems, those containing axial symmetry, which are somewhat analogous to two-dimensional problems, where the fluid properties in the ϕ-direction are constant.

From the conservation of mass, the mass flow rate through any spherical surface area, $4\pi R^2$, enclosing the source must be constant. If we designate the fluid constant density by the symbol ρ, we have the following equation from the conservation of mass:

$$Q\rho = 4\pi R^2 \, v_R \, \rho \qquad\qquad (3\text{-}8)$$

or

$$v_R = \frac{Q}{4\pi R^2} \qquad\qquad (3\text{-}9)$$

where the velocity of fluid in the radial direction, v_R, is a function of the distance, R, from the center of the source. Note that at the center, where R = 0, the radial velocity has an infinite magnitude. Thus, while this source is a useful concept, it cannot exactly duplicate physical fluid flow phenomena at its center.

The velocity components in the θ and ϕ directions, v_θ and v_ϕ are zero.

$$v_\theta = 0 \qquad\qquad (3\text{-}10)$$

$$v_\phi = 0 \qquad\qquad (3\text{-}11)$$

In order to determine the velocity potential, Φ, for this flow in spherical coordinates (R, θ, ϕ), equation (3-5) becomes

$$\mathbf{V} = \boldsymbol{\nabla}\Phi = \partial_R \Phi \, \hat{\mathbf{R}} + \frac{1}{R}\partial_\theta \Phi \, \hat{\boldsymbol{\theta}} + \frac{1}{R\sin\theta}\partial_\phi \Phi \, \hat{\boldsymbol{\phi}} \qquad\qquad (3\text{-}12)$$

where I use the partial differential notation $\partial_x \equiv \frac{\partial}{\partial x}$, and a hat on a symbol indicates the unit vector in the direction of the symbol's axis.

Equating the velocity components in equations (3-9) through (3-11) to those components in equation (3-12) results in

$$\partial_R \Phi = \frac{Q}{4\pi R^2} \tag{3-13}$$

$$\frac{1}{R}\partial_\theta \Phi = 0 \tag{3-14}$$

$$\frac{1}{R\sin\theta}\partial_\phi \Phi = 0 \tag{3-15}$$

The velocity potential for the three-dimensional incompressible source can be determined by integrating these three equations, and equating their components to solve for Φ. By carrying out those calculations, the source velocity potential becomes

$$\Phi_{source} = -\frac{Q}{4\pi R} + C_1 \tag{3-16}$$

where C_1 is an arbitrary constant of integration. Similarly, it can be shown that the velocity potential for the three-dimensional sink is

$$\Phi_{sink} = \frac{Q}{4\pi R} + C_2 \tag{3-17}$$

It can also be easily shown that the source and sink velocity potentials satisfy the Laplace equation simply by substitution. In the analysis to follow, the integration constants will not be of any consequence, so they will be omitted.

Combined Source and Sink Field

Since our objective is to determine the force between an incompressible fluid source and sink due to their combined flow field, we will now derive their flow field. Consider a three-dimensional incompressible fluid flow source located at the origin, and an incompressible fluid flow sink with equal strength located on the reference z-axis at z = a in the spherical coordinate system as shown in Figure 3-2.

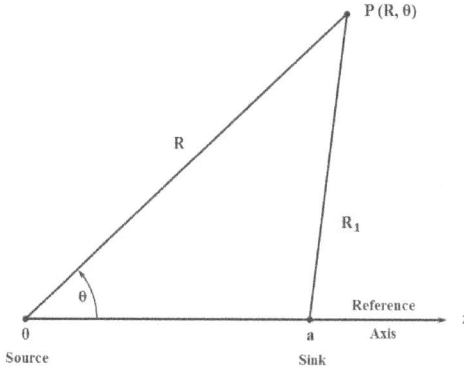

Figure 3-2. Geometry for source-sink flow.

Ignoring the constants, the velocity potential of the source at the origin for any point P(R, θ) is given by equation (3-16) to be

$$\Phi_{source} = - \frac{Q}{4 \pi R} \qquad\qquad (3\text{-}18)$$

and that of the sink at z = a is

$$\Phi_{sink} = \frac{Q}{4 \pi R_1} \qquad\qquad (3\text{-}19)$$

Since both Φ_{source} and Φ_{sink} satisfy the Laplace equation, their sum also satisfies the Laplace equation. Thus,

$$\Phi = - \frac{Q}{4 \pi R} + \frac{Q}{4 \pi R_1} \qquad\qquad (3\text{-}20)$$

represents the velocity potential of the combined source-sink flow field.

Now, since the sink is offset from the origin, we will need to replace R_1 in terms of R and θ. For the location at $z = a$, the sink's velocity potential becomes

$$\Phi_{sink} = \frac{Q}{4\pi\left(R^2 + a^2 - 2\,a\,R\cos\theta\right)^{1/2}} \tag{3-21}$$

Therefore, the combined velocity potential for the source-sink field becomes

$$\Phi = -\frac{Q}{4\pi R} + \frac{Q}{4\pi\left(R^2 + a^2 - 2\,a\,R\cos\theta\right)^{1/2}} \tag{3-22}$$

Derivation of Force Between an Incompressible Source and Sink

To determine the force on the source due to the presence of the sink we will construct a fixed spherical control surface of radius, r, around the source as shown in Figure 3-3, and calculate the force on the fluid within the control volume.

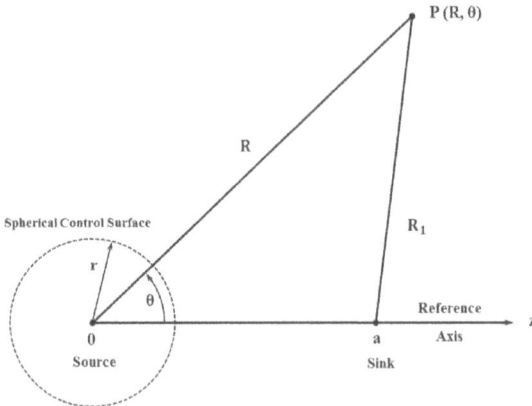

Figure 3-3. Geometry for spherical control surface around a source

To determine this force, we will use the momentum equation for a control volume in an inertial reference frame [25, p.67, (2-24)],

$$\mathbf{F} = \oiint \mathbf{V}(\rho \mathbf{V} \cdot d\mathbf{A}) + \frac{\partial}{\partial t} \iiint \mathbf{V} \rho \, dV \tag{3-23}$$

where $\mathbf{F}$ is the resultant of all body forces and surface forces exerted by the surround
ings on the fluid within the control volume, $\mathbf{V}$ is the fluid velocity, ρ is the fluid
density, $d\mathbf{A}$ is the surface area vector of the control surface, and dV is a fluid
volume element inside the control volume.

Under steady state conditions, the rate of change of momentum inside the control
volume is zero, and the equation becomes

$$\mathbf{F} = \oiint \mathbf{V}(\rho \mathbf{V} \cdot d\mathbf{A}) \tag{3-24}$$

That is, the force, $\mathbf{F}$, equals the integral of the momentum flux over the surface of
the control volume.

In general fluids engineering practice this equation is useful, for example, to
determine the forces required to hold pipes in place containing moving fluids, the
forces involved in jet propulsion, hydrodynamic turbines, and numerous other
applications. This is achieved by enclosing the fluid with a control surface at the
physical flow boundaries constraining the fluid, and at the inlet and outlet surfaces
through which fluid flows. In many of these practical cases, fluid properties are
approximated by averaging them over the various inlet and outlet flow surfaces in
order to estimate a resultant force exerted by or on the moving fluid. Now, let's
consider a short simplified practical example that has some similarity to the problem we will be analyzing involving the source and sink.

Rocket Motor Example

In Figure 3-4, we have a rocket motor that is fixed on a test stand with some
combustible material inside that has been ignited, so we see high velocity exhaust
being forced out of the nozzle.

Normally, we might be interested in determining the force required to constrain the
rocket motor on the test stand or the thrust on the rocket motor. For this simple

example, we are interested in determining the force, **F**, on the fluid inside the control volume at this particular instant under these particular conditions.

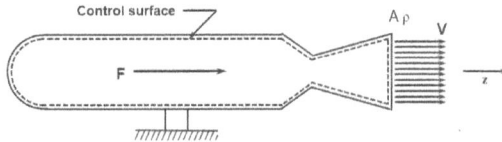

Figure 3-4. A Rocket Motor on a Test Stand

We have enclosed the fluid inside the rocket motor with a control surface to determine the force on the fluid within it. For the integration calculation, we only have
to be concerned with the control volume surface at the nozzle end because that is
the only surface area on the control volume which has momentum flux through it.
Given the surface area of the exhaust nozzle, A, the average velocity of the exhaust
particles, **V**, and the average density of the exiting exhaust, ρ, we can calculate the
force on the fluid within the control volume. For this case, the force will be in the z-
direction, and the applicable equation (3-24) becomes

$$\mathbf{F}_z = \oiint \mathbf{V}_z(\rho \mathbf{V} \cdot d\mathbf{A}) \qquad\qquad (3\text{-}25)$$

where $\mathbf{V}_z$ is the velocity vector in the z-direction. There are two sign conventions
for this equation. The signs of $\mathbf{F}_z$ and $\mathbf{V}_z$ depend on the positive direction chosen
for the coordinate z-axis, and the sign for the dot product $(\rho \mathbf{V} \cdot d\mathbf{A})$ depends on the
local orientation of the fluid velocity vector, **V**, relative to the surface area vector,
$d\mathbf{A}$. The surface area vector is always perpendicular to the surface and pointing
away from the control volume.

In this simple case the quantities and orientations are such that the signs are all positive. Integrating over the entire control surface, we can immediately write down the force as

$$\mathbf{F} = V(\rho \, V \, A)\, \hat{z} = \rho V^2 A \, \hat{z}$$

This is the force on the fluid inside the control volume, which causes the fluid to move to the right, and it is also the force required to constrain the rocket motor from flying off the stand. They are both in the positive z-direction.

A Net Force Caused by Asymmetry

In the analogy to the source and sink problem, the source is somewhat like the rocket motor in that it has something internal causing particles to be streaming out, and we have constrained it to stay at the origin. But, unlike the rocket motor the particles of a source are streaming radially in all directions. If the source were isolated by itself without the sink, the momentum vectors of the particles would be symmetric about the origin, and there would be no net force on the control volume, but when we introduce the sink into the source's field, the entire source flow field changes. It is hypothesized that it is the introduction of the sink that causes the field around the source to be unsymmetrical in the z-direction, and creates a net force on the control volume. Let's formalize that hypothesis, and moreover do it in terms of electrical point charges.

Hypothesis II:

It is the introduction of a point charge into the field of another point charge that causes the fields around each to become unsymmetrical, which causes a net force between them.

For the source and sink case to be developed here, the approximations made in the rocket motor example by averaging the velocity and density will not be necessary. The equation shall be determined *exactly* because we will know the exact functions involved in the calculation. As in the example, we are particularly interested in the force acting on the fluid within the control surface in the direction of the reference z-axis. Also, the combined flow field will be axially symmetric about the z-axis, so there will be no net force perpendicular to it.

Now that we have all the necessary ingredients, we will determine the components of equation (3-25) to prepare for the integration. Equations (3-12) and (3-22) can be used to determine the velocity vector function for the combined fluid flow field. Since the combined flow field is axially symmetric, we will only have contributions to the velocity components in the R and θ directions. Those calculations can be shown to result in the velocity vector

$$\mathbf{V} = [\frac{Q}{4\pi R^2} - \frac{Q(R - a\cos\theta)}{4\pi(R^2 + a^2 - 2aR\cos\theta)^{3/2}}]\,\hat{\mathbf{R}} - \frac{Qa\sin\theta}{4\pi(R^2 + a^2 - 2aR\cos\theta)^{3/2}}\,\hat{\theta} \tag{3-26}$$

The relative velocities associated with this flow are plotted in Figure 3-5. Infinite velocities at the source and sink centers cannot be drawn, but the gist of the velocity field can be seen. The magnitudes of the velocities decrease rapidly from the centers, and are difficult to see away from the singularities when drawn to scale. The density field which is not shown would be easy to plot. Since it is a constant value, no pattern would be visible. Of course, a constant density world cannot model reality because there would be no concentrations of mass that we could call matter.

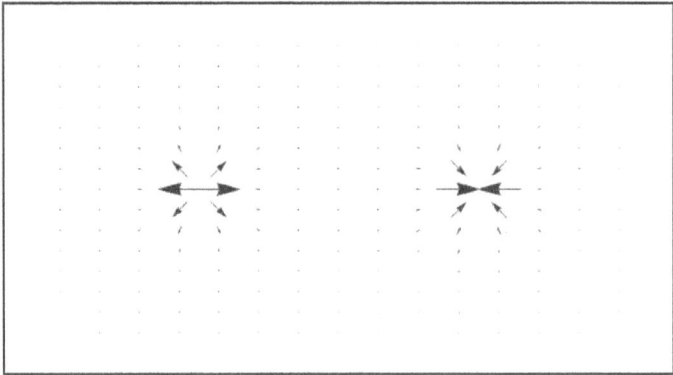

Figure 3-5. Velocity vector plot of an incompressible source-sink fluid field.

To prepare for the integration over the spherical control surface with constant radius, r, we must replace R in equation (3-26) with r, resulting in the velocity at the control surface

$$\mathbf{V}_{cs} = [\frac{Q}{4\pi r^2} - \frac{Q(r - a\cos\theta)}{4\pi(r^2 + a^2 - 2ar\cos\theta)^{3/2}}]\,\hat{\mathbf{R}} - \frac{Qa\sin\theta}{4\pi(r^2 + a^2 - 2ar\cos\theta)^{3/2}}\,\hat{\boldsymbol{\theta}} \qquad (3\text{-}27)$$

To determine $\mathbf{V}_z$ on the control surface, we can substitute in place of the unit vectors $\hat{\mathbf{R}}$ and $\hat{\boldsymbol{\theta}}$ their z-components

$$\hat{\mathbf{R}}_z = \cos\theta\,\hat{z} \qquad (3\text{-}28)$$

$$\hat{\boldsymbol{\theta}}_z = -\sin\theta\,\hat{z} \qquad (3\text{-}29)$$

The result becomes

$$\mathbf{V}_z = ([\frac{Q}{4\pi r^2} - \frac{Q(r - a\cos\theta)}{4\pi(r^2 + a^2 - 2ar\cos\theta)^{3/2}}]\cos\theta + \frac{Qa\sin\theta}{4\pi(r^2 + a^2 - 2ar\cos\theta)^{3/2}}\sin\theta)\,\hat{z} \qquad (3\text{-}30)$$

The differential surface area of the spherical control surface of radius r over which we shall integrate is

$$d\mathbf{A} = 2\pi r^2 \sin\theta\, d\theta\, \hat{\mathbf{R}} \qquad (3\text{-}31)$$

Calculating the dot product in equation (3-25), we have

$$(\rho\mathbf{V}_{cs} \cdot d\mathbf{A}) = 2\pi\rho r^2 \sin\theta\,[\frac{Q}{4\pi r^2} - \frac{Q(r - a\cos\theta)}{4\pi(r^2 + a^2 - 2ar\cos\theta)^{3/2}}]\, d\theta \qquad (3\text{-}32)$$

Now, substituting all of these derived quantities into equation (3-25), we find the force on the control volume to be

$$\mathbf{F}_z = 2\pi\rho r^2 \int_0^\pi \sin\theta \left[\frac{Q}{4\pi r^2} - \frac{Q(r - a\cos\theta)}{4\pi(r^2 + a^2 - 2ar\cos\theta)^{3/2}}\right]$$

$$(\left[\frac{Q}{4\pi r^2} - \frac{Q(r - a\cos\theta)}{4\pi(r^2 + a^2 - 2ar\cos\theta)^{3/2}}\right]\cos\theta + \frac{Qa\sin\theta}{4\pi(r^2 + a^2 - 2ar\cos\theta)^{3/2}}\sin\theta\,)\,d\theta\,\hat{z} \quad (3\text{-}33)$$

This equation can be simplified to result in the following:

$$\mathbf{F}_z = \frac{Q^2\rho}{8\pi r^2}\int_0^\pi \left((r^3 - ar^2\cos\theta - (r^2 + a^2 - 2ar\cos\theta)^{3/2}\right)$$
$$\left(-ar^2 + r^3\cos\theta - \cos\theta\,(r^2 + a^2 - 2ar\cos\theta)^{3/2}\right)$$
$$\sin\theta\,d\theta)/(r^2 + a^2 - 2ar\cos\theta)^3$$

$$\hat{z} \qquad (3\text{-}34)$$

This equation was integrated using the Wolfram Research *Mathematica* integrator, and since the problem of interest is to find the force on an incompressible source at the origin the size of a positron, the radius of the spherical control volume was set to the radius of a positron. The main results follow in the next two plots which show the forces from both the fluid data and Coulomb's Law vs. separation distance in different regions of separation between the source and sink.

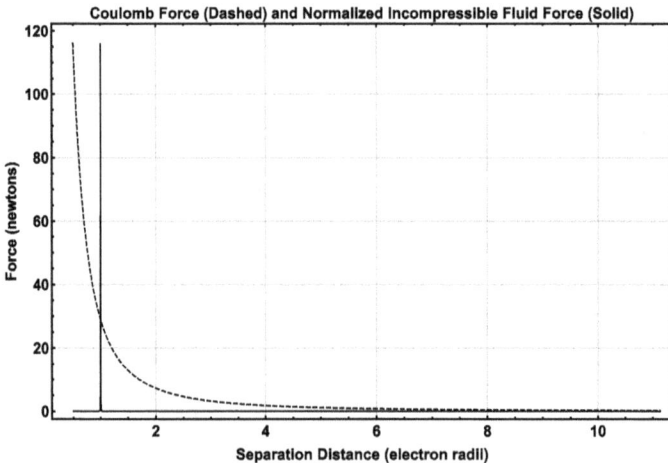

Figure 3-6. The force $\mathbf{F}_z$ on the fluid within the control volume (Solid) normalized to the maximum of the plotted classical electrostatic force curve (Dashed) versus values of a/r.

Plotted in Figure 3-6 are the classical electrostatic Coulomb force between an electron and a positron indicated by a dashed line, and a fluid force curve which is solid. To be able to visualize both force curves on the same plot the fluid force needed to be normalized. To normalize the fluid force the value of the volume flow rate, Q, was adjusted so the maximum fluid force equaled the maximum force of the electrostatic curve. The value of the fluid force was not calculated at exactly a/r = 1 because at that location the control surface crosses the origin of the sink which is a singular point. This causes the fluid force to be theoretically infinite, and unplotable. Only the remainder of the fluid force curves on each side of a/r = 1 are shown. We can see that the fluid force increases abruptly from both sides of the abscissa at a/r = 1. Because large values of force were normalized on each side of a/r = 1, the rest of the fluid force curves on each side appear to be nearly zero, and don't show much of their real characteristics. Of course, the electrostatic Coulomb force curve also approaches infinity as a/r approaches zero.

In Figure 3-7, I have plotted the two curves over a range where the separation distance is about 2 ≤ a/r < 8, in which the spherical control surface around the source does not cross the sink's center. This range has eliminated both the Coulomb and fluid force infinities that are physically unrealistic. I have again adjusted the source and sink volume flow rates, Q, but instead of normalizing to a value of the Coulomb force on the left side of the plot, this time I adjusted it so the force curves match at a large separation distance slightly less than a/r = 54, way off the chart to the right. Note that both force curves are always positive indicating an attractive force, and now match fairly closely, increasing in magnitude with decreasing separation distance. However, they have visibly different rates of climb when the separation distance is very small. The fluid force increases a little more rapidly than the electrostatic force.

In this analysis I assumed the density, ρ, of the entire source and sink flow field to be constant throughout, and equal to the average mass density of an electron. Under these conditions the volume flow rate is approximately

$$Q = 1.496 \times 10^{-20} \text{ m}^3/\text{sec} \tag{3-35}$$

and the equivalent mass flow rate is then

$Q \rho = 1.454 \times 10^{-7} \text{ kg/sec.}$ (3-36)

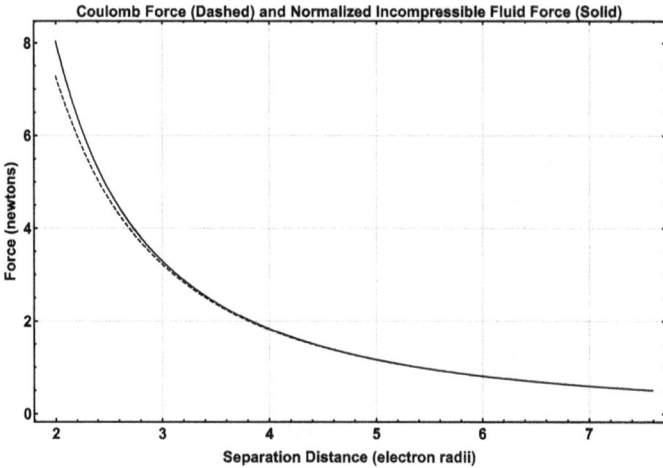

Figure 3-7. The force F_z on the fluid within the control volume (Solid) normalized to the classical electrostatic force curve (Dashed) at about $a/r = 54$, versus values of a/r approximately in the region $2 \leq a/r < 8$.

The numerical data used in the calculations above came from the following 2010 CODATA [26].

$e = 1.602176565 \times 10^{-19} \text{ coulomb}$
$r = 2.8179403267 \times 10^{-15} \text{ m}$
$m_e = 9.10938291 \times 10^{-31} \text{ kg}$
$c = 2.99792458 \times 10^8 \text{ m/sec}$
$\epsilon_0 = 8.854187817 \times 10^{-12} \text{ coulomb}^2/\text{nt-m}^2$

Even though the calculation method and singularities used above allowed me to derive the exact fluid force on a control volume which closely matches Coulomb's Law for separation distances of about $a/r \gtrsim 3$, the incompressible source and sink produce fluid flow characteristics that are completely inconsistent with electromagnetic phenomena.

If electrons and positrons were constant density sources and sinks, they would have the same density of the surrounding fluid, there would be no high concentrations anywhere within the fluid field which we could call massive objects, and electrons and positrons would be visibly indistinguishable from the rest of the field. Also, the velocity is infinite at the centers of the source and sink, exceeding the maximum speed of light. In fact, incompressible fluids and rigid bodies are physically impossible according to special relativity. A disturbance in these objects would be transmitted through them instantaneously.

In the derivation of the fluid flow above, I used the classical velocity potential function. When this velocity potential is used to develop a more general theory of a *compressible* fluid, the resulting equation is nonlinear and almost a hopeless task to solve. To help eliminate this problem, the next step, which is normally pursued, is to find an approximate solution by using a method such as *small-perturbation theory* to linearize the equation. Fortunately, since the interest here is to *exactly* connect fluid dynamics principles and electromagnetic phenomena, I was able to find the key for developing a new potential function to open the door to this relation- ship.

In the next chapter, I shall construct a derivation nearly identical to that done above, except I will instead develop a *compressible* fluid source, sink, and generalized potential function that will produce an exact nonperturbative compressible theory and solution. These new functions will allow the reversal of the roles of velocity and density, i.e. they allow the magnitude of velocity to be a constant and the density to vary throughout the flow field. This will result in a fluid flow theory of the electron and positron with physical attributes *precisely* matching their experimen tally determined characteristics and values.

Chapter 4

Theory of the Electron and Positron (and everything else)

James Clerk Maxwell, Hendrik Antoon Lorentz, as well as many other researchers and authors were correct when they thought that electromagnetic phenomena are in some ways analogous to the continuous flow of fluids. However, most authors have also been careful to state that this analogy is not exactly correct, stating that nothing is actually flowing in an electric field. In this chapter, this frequent statement will be refuted. Also, it will be shown that the true fluid analogy is not ideal *incompressible* fluid flow, as sometimes referred to as being analogous, but is *compressible* fluid flow. Putting this in the form of a hypothesis:

Hypothesis III:

The fundamental basis of electromagnetic phenomena is compressible fluid flow.

Although the mathematics in the theory to be derived below, concerning compressible fluid flow, will be describing a source and a sink to represent a positron and electron using continuum mechanics principles, the outward and inward fluid flows will actually not be continuous, but will be assumed to be composed of very small particles of mass. This is consistent with all fluids. On a microscopic scale most fluids can be thought to consist of molecules or atoms. However, the fluid flows originating from and terminating at point charges must consist of even smaller particles, much much smaller than an electron. Now, since the mass "objects" in Hypothesis I are believed to move between point charges to transmit the electric force, they must constitute the electric field, and move at the speed of light in a vacuum. If a particle with infinitesimal mass such as this exists in nature, according to the theory of special relativity it must have zero rest mass. This may make no sense from a classical point of view, but is admissible in relativity theory. Therefore, I shall postulate the necessary existence of a new quantum particle of mass to replace the "object" in Hypothesis I, which I shall call an *electron dust particle*, and abbreviate the name of these particles as *edust particles* or just *edust*.

Postulate I:

There exists a fundamental sub-electronic quantum particle which has mass and moves at the speed of light in a vacuum (which shall be called an electron dust particle or edust for short).

Combining this postulate with Hypothesis I, we have:

Hypothesis IV:

The electric fields of point charges, i.e. positrons and electrons, consist of edust streaming either away from or toward their centers in all radial directions. It is unknown at this time whether edust particles stream away from or toward positrons, but whichever applies the opposite is true for electrons.

In 1930, Wolfgang Pauli invented a new particle to explain an energy deficit in beta decay. In doing so, he apparently exclaimed that he had done a "terrible thing," stating that "I have postulated a particle that cannot be detected." [37, p.152] Although Pauli's particle, the neutrino, was apparently eventually detected, I have my doubts that this will ever occur for any individual edust particle because Pauli's neutrino was very difficult to detect, and according to the theory that will evolve here the neutrino is expected to be composed of a concentration of many edust particles. On the other hand, this theory will conclude that all matter is composed *solely* of edust. Thus, in a way, edust has already been detected in the form of matter everywhere. But, I'm getting ahead of myself again. So, let's continue to build the necessary foundation for the theory.

Now, it is well known that the speed of propagation of light in a vacuum is a constant. However, this is only a *free-field* electromagnetic phenomenon. That is, the speed of light is slower than its maximum when it passes through matter, and varies with the density of matter. Now, I have postulated that electron dust particles move at the speed of light in a vacuum. However, in the theory that is to be developed, the speed of edust particles must *always* be the maximum speed of light regardless of the density of matter through which they may propagate, or with which they collide, except during collisions. To formalize this, I will put it in the form of a postulate before beginning the derivation of the equations governing edust continuum flow.

Postulate II:

While not engaged in a collision, the speed of electron dust particles is <u>always</u> the maximum speed of light.

To simplify the rest of the text, from now on I will not always add "in a vacuum" or "maximum" to the phrase "speed of light" when referring to edust particles, but assume their speed is always associated with the value of the maximum speed of light, c.

Assuming the conservation of momentum is valid in the flow of edust, this postulate has several significant consequences. First, if edust particles always move at the maximum speed of light except during collisions, then it follows that they must be perfectly elastic and spherical. This would not be possible for a one-dimensional oscillating line or loop string, or any other oddly shaped object. Second, they must not be subject to any other forces except collisions, since those forces would affect their speed. And third, edust particles may *only* collide with themselves, or possibly stationary bodies with infinite mass, if they were to exist. For now, we will assume there are no stationary bodies with infinite mass in our universe, except perhaps at a boundary, if there is a boundary.

It follows from the discussion above that since edust particles are not subject to any forces other than collisions, then they have no electric charge, and even though they have mass, they are not subject to the force of gravity. It also follows that electron dust particles are the *only* elementary particles in the universe. Therefore, if this theory proves to be true, gravity must be a phenomenon *caused* by colliding electron dust particles. Also, all other objects, e.g. protons, neutrons, electrons, positrons, neutrinos, photons, etc., are composite particles composed solely of flowing electron dust particles. Furthermore, massless particles which carry momentum do not exist.

Now, in order to summarize and proceed, I shall propose the hypothesis below which combines the main ideas in the previous postulates and hypotheses.

Hypothesis V:

Point charges, i.e. positrons and electrons, are sources and sinks that constantly either emit or absorb electron dust particles which move at the speed c in all radial directions to produce variable mass density flow patterns, commonly called electric fields. It is unknown at this time whether edust particles stream away from or toward positrons, but whichever applies the opposite is true for electrons.

To test the above postulates and hypotheses, I will now introduce new compressible source and sink velocity and density functions, and a new scalar potential function valid for *both* incompressible and compressible fluids. Later, in opposition to setting the density to be a constant and solving for a velocity function as done previously in Chapter 3, I will use the new scalar potential function for a special compressible flow by making the unusual fluid field assumption that *the fluid speed is constant everywhere,* and allow the density to vary. First, let's begin with the derivation of a new generalized potential function.

A New Potential Function - the Momentum Density Potential

I will now derive a potential function compatible with *both* incompressible and compressible fluid flow, and call it the Momentum Density Potential. It is a generalization of the velocity potential used in incompressible fluid flow problems. Here we will *not* impose any requirements on the density of our fluid flow fields as done previously, but use the partial differential equation of continuity (3-2) which is valid for *any steady* fluid flow.

$$\nabla \cdot \rho \mathbf{V} = 0 \qquad\qquad (3\text{-}2)$$

Again, I will also require the flow to be irrotational, so that the **curl** of the momentum density vector, $\rho\mathbf{V}$, in the fluid flow fields is equal to zero.

$$\nabla \times \rho\mathbf{V} = 0 \tag{4-1}$$

Note that equation (4-1) implicitly requires the condition of irrotational flow. Since $\rho\mathbf{V}$ is collinear with the vector $\mathbf{V}$, if $\nabla \times \rho\mathbf{V} = 0$, then the condition of irrotational flow, $\nabla \times \mathbf{V} = 0$ is also satisfied.

Now, as above for incompressible flow, if the **curl** of a vector is everywhere zero, then that vector can be expressed as the gradient of some scalar potential. Assuming irrotational flow, a point function called the Momentum Density Potential which I will denote by the symbol mho, $\mho$, can be defined. Although $\mho$ is used in other books for other quantities, its use is rare and those quantities will not be used at all in this book. Then,

$$\rho\mathbf{V} = \nabla\mho \tag{4-2}$$

Now substituting from equation (4-2) into equation (3-2), we get

$$\nabla \cdot \nabla\mho = 0 \tag{4-3}$$

and our Laplace equation for *any* steady irrotational fluid flow becomes

$$\nabla^2\mho = 0 \tag{4-4}$$

This is a Laplace equation compatible with both incompressible and compressible fluid flow. In the next section, I shall derive the desired *compressible* source and sink flow fields that satisfy this equation. Later, since the Laplace equation is a linear equation, we will have the freedom to add these compressible solutions to form a new compressible fluid flow field containing both of them. These source and sink flow fields will have *exactly* the characteristics of electron and positron electrostatic fields.

A Three-Dimensional Compressible Source and Sink

In the field of fluid dynamics, only *incompressible* sources and sinks have been used, to the best of my knowledge, in the construction of potential fluid flow fields. Potential compressible sources and sinks have not been previously conceived or developed. I will derive a compressible, meaning variable density, fluid source and sink here. The motivation is to show that the underlying physics of compressible fluid flow field theory has results that match *exactly* that of electrostatic field theory. That is, the developed compressible flow fields which exhibit a flow velocity that is constant everywhere, and have a mass density that varies with distance from the center of sources and sinks, are completely consistent with electron and positron electrostatic field phenomena.

I shall begin by defining a steady compressible three-dimensional fluid source at the origin of a fixed inertial reference frame with spherical coordinates as shown in Figure 4-1.

Previously, I postulated that edust particles always travel at the maximum speed of light, c, and for this flow their velocities are in the radial direction: positive for a source, and negative for a sink. From the conservation of mass in the field, the mass flux of edust particles through any spherical control surface area, $4\pi R^2$, enclosing a source or sink, must be constant. If we let the mass flux constant be designated by the letter K, we have

$$K = 4\pi R^2\, c\, \rho \qquad\qquad (4\text{-}5)$$

or

$$\rho = \frac{K}{4\pi c R^2} \qquad\qquad (4\text{-}6)$$

where ρ is now the density of the edust fields for both sources and sinks, which is a function of the distance, R, from their centers.

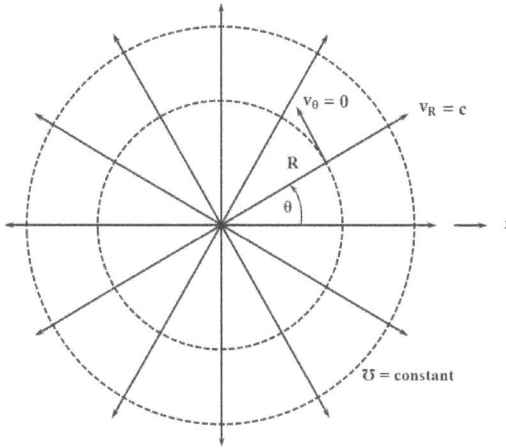

Figure 4-1 A steady three-dimensional compressible fluid source

The vector form of the velocity field, **V**, in spherical coordinates for this type of compressible source is

$$\mathbf{V} = c\,\hat{\mathbf{R}} = \frac{K}{4\pi\rho R^2}\,\hat{\mathbf{R}} \tag{4-7}$$

where $\hat{\mathbf{R}}$ is the unit vector in the radial direction. The velocity vector simply has a negative sign for a sink.

A Possible Case of Fluid Flow

Since the last section describes a new type of source and sink, we will first check to see if they are possible. Any mathematical function which represents a possible case of steady flow must satisfy the differential equation of continuity repeated here.

$$\nabla \cdot \rho \mathbf{V} = 0 \tag{3-2}$$

where ∇ is the vector gradient operator, and $\mathbf{V}$ is the velocity vector. In the spherical coordinate system (R, θ, ϕ),

$$\mathbf{V} = v_R \,\hat{\mathbf{R}} + v_\theta \,\hat{\boldsymbol{\theta}} + v_\phi \,\hat{\boldsymbol{\phi}} \tag{4-8}$$

and the scalar form of equation (3-2) becomes

$$\frac{1}{R^2}\partial_R\left(R^2\,\rho\,v_R\right) + \frac{1}{R\sin\theta}\partial_\theta(\rho\,v_\theta\sin\theta) + \frac{1}{R\sin\theta}\partial_\phi(\rho\,v_\phi) = 0 \tag{4-9}$$

For the steady compressible source derived above

$$v_R = c \tag{4-10}$$

$$v_\theta = 0 \tag{4-11}$$

$$v_\phi = 0 \tag{4-12}$$

and

$$\rho = \frac{K}{4\pi c R^2} \tag{4-13}$$

Substituting these quantities into equation (4-9), we find immediately that

$$\frac{1}{R^2}\partial_R\left(\frac{K}{4\pi}\right) + 0 + 0 = 0 \tag{4-14}$$

Now, since the partial derivative of a constant is zero,

$$0 = 0 \tag{4-15}$$

We see that the equation is satisfied. This is also true for the sink where $v_R = -c$.

Satisfying the Curl of the Momentum Density Vector

We will now show that the **curl** of the momentum density vector of our compressible source is zero. That is

$$\nabla \times \rho \mathbf{V} = 0 \tag{4-1}$$

The form of equation (4-1) expanded in spherical coordinates is

$$\frac{1}{R \sin \theta}[\partial_\theta \ (\rho \ v_\phi \sin \theta) - \partial_\phi \ (\rho \ v_\theta)] \ \hat{\mathbf{R}}$$
$$+ \frac{1}{R}[\frac{1}{\sin \theta} \ \partial_\phi \ (\rho \ v_R) - \partial_R \ (R \rho \ v_\phi)] \ \hat{\boldsymbol{\theta}}$$
$$+ \frac{1}{R}[\ \partial_R \ (R \rho \ v_\theta) - \partial_\theta \ (\rho \ v_R)] \ \hat{\boldsymbol{\phi}} = 0 \tag{4-16}$$

Substituting from equations (4-10) through (4-12) into equation (4-16), we can easily see that since two of the velocity components are zero, four of the six terms drop out immediately, and we are left with

$$\frac{1}{R}[\frac{1}{\sin \theta} \ \partial_\phi \ (\rho \ c)] \ \hat{\boldsymbol{\theta}} - \frac{1}{R}[\partial_\theta \ (\rho \ c)] \ \hat{\boldsymbol{\phi}} = 0 \tag{4-17}$$

But, it is also easy to see that ρ and c are not functions of ϕ or θ. Thus, the partial derivatives of their product are zero, and the equation is satisfied, $0 = 0$. This is similarly true for the sink.

Momentum Density Potentials of a Compressible Source and Sink

Now that I have shown that Momentum Density Potentials exist for the defined compressible fluid flow source and sink, we will derive the potential here for the source, then easily conclude the form for the sink. From equation (4-2),

$$\rho \mathbf{V} = \nabla \mho \tag{4-2}$$

or in expanded form,

$$\rho \, v_R \, \bar{\hat{\mathbf{R}}} + \rho \, v_\theta \, \hat{\boldsymbol{\theta}} + \rho \, v_\phi \, \hat{\boldsymbol{\phi}} = \partial_R \, \mho \, \hat{\mathbf{R}} + \frac{1}{R}\partial_\theta \, \mho \, \hat{\boldsymbol{\theta}} + \frac{1}{R \sin \theta}\partial_\phi \, \mho \, \hat{\boldsymbol{\phi}} \tag{4-18}$$

Equating the corresponding vector component quantities, we have

$$\rho \, v_R = \partial_R \, \mho \tag{4-19}$$

$$\rho \, v_\theta = \frac{1}{R}\partial_\theta \, \mho \tag{4-20}$$

$$\rho \, v_\phi = \frac{1}{R \sin \theta}\partial_\phi \, \mho \tag{4-21}$$

Inserting the velocity components and density from equations (4-10) through (4-13), we have

$$\partial_R \, \mho = \frac{K}{4 \pi R^2} \tag{4-22}$$

$$\partial_\theta \, \mho = 0 \tag{4-23}$$

$$\partial_\phi \, \mho = 0 \tag{4-24}$$

The desired Momentum Density Potential for the three-dimensional source of compressible fluid flow can be determined by integrating these three equations and equating their components. From this, the Momentum Density Potential becomes

$$\mho_{\text{source}} = -\frac{K}{4 \pi R} \tag{4-25}$$

where the arbitrary constant of integration has been omitted. Similarly, it can easily be shown that the Momentum Density Potential for the sink is

$$\mho_{\text{sink}} = \frac{K}{4 \pi R} \tag{4-26}$$

Notice the similarity between these last two equations and equations (3-16) and (3-17) if their constants of integration are omitted. They are identical in form. The only difference is that the velocity potentials are proportional to the volume flow rate, Q, while the Momentum Density Potentials are proportional to the mass flow rate, K.

Combined Compressible Source and Sink Field

It can be easily shown that the derived source and sink Momentum Density Potentials satisfy the Laplace equation by substitution. Therefore, their sum also satisfies it. Since our objective is to determine the force between a compressible source and sink due to their combined flow field, we will now derive the potential equation characterizing the flow field. Consider a three-dimensional *compressible* source located at the origin, and the *compressible* sink with equal mass flow rate located on the reference z-axis at z = a as in Figure 4-2.

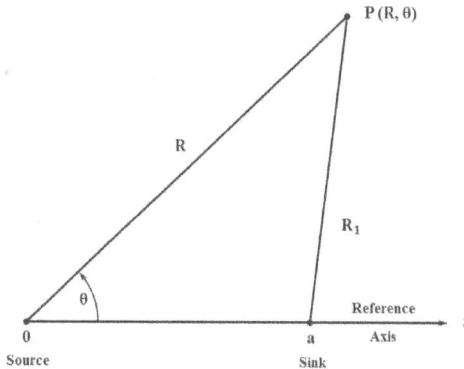

Figure 4-2. Geometry for compressible source-sink fluid flow.

The Momentum Density Potential of the source at the origin is given by equation (4-25) to be

$$\mho_{source} = -\frac{K}{4\pi R} \tag{4-25}$$

and that of the sink at z = a is

$$\mho_{sink} = \frac{K}{4\pi R_1} \tag{4-27}$$

Since both $\mho_{source}$ and $\mho_{sink}$ each satisfy the Laplace equation, their sum

$$\mho = -\frac{K}{4\pi R} + \frac{K}{4\pi R_1} \tag{4-28}$$

also satisfies it.

As required previously, since the sink is offset from the origin, we will need to replace R_1 in terms of R and θ. For the location at z = a, the sink's Momentum Density Potential becomes

$$\mho_{sink} = \frac{K}{4\pi(R^2 + a^2 - 2aR\cos\theta)^{1/2}} \tag{4-29}$$

Therefore, the combined Momentum Density Potential for the compressible source-sink field becomes

$$\mho = -\frac{K}{4\pi R} + \frac{K}{4\pi(R^2 + a^2 - 2aR\cos\theta)^{1/2}} \tag{4-30}$$

Derivation of the Force Between a Compressible Source and Sink

To determine the force on the source due to the presence of the sink consider the spherical control surface in Figure 4-3. As done previously, I will calculate the force on the spherical control volume in the direction of the reference z-axis using equation (3-25).

$$F_z = \oiint V_z(\rho V \cdot dA) \tag{3-25}$$

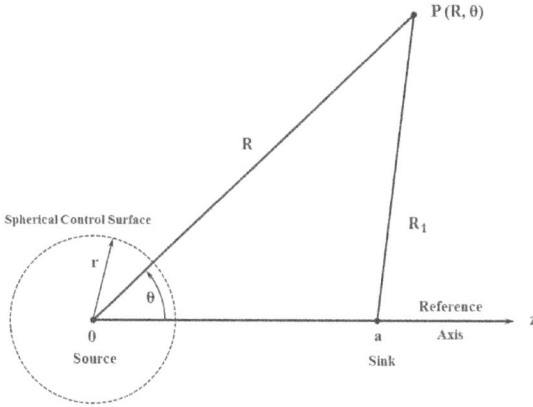

Figure 4-3. Geometry for spherical control surface around a source

I will now determine the components of this equation. Equation (4-18), repeated below, can be used to determine the velocity vector components for the combined fluid flow field from the Momentum Density Potential (4-30), also repeated below.

$$\rho \, v_R \, \hat{\mathbf{R}} + \rho \, v_\theta \hat{\theta} + \rho \, v_\phi \hat{\phi} = \partial_R \mho \, \hat{\mathbf{R}} + \frac{1}{R}\partial_\theta \mho \, \hat{\theta} + \frac{1}{R \sin \theta}\partial_\phi \mho \, \hat{\phi} \tag{4-18}$$

$$\mho = - \frac{K}{4\pi R} + \frac{K}{4\pi\left(R^2 + a^2 - 2\,a\,R\cos\theta\right)^{1/2}} \tag{4-30}$$

The combined fluid flow field will be axially symmetric, so we will not have any velocity component in the ϕ direction. Inserting $\mho$ into equation (4-18) and solving for the vector components results in the velocity vector

$$\mathbf{V} = \frac{1}{\rho}\left[\frac{K}{4\pi R^2} - \frac{K\,(R - a\cos\theta)}{4\pi\left(R^2 + a^2 - 2\,a\,R\cos\theta\right)^{3/2}}\right] \hat{\mathbf{R}} - \frac{1}{\rho}\frac{K\,a\sin\theta}{4\pi\left(R^2 + a^2 - 2\,a\,R\cos\theta\right)^{3/2}} \hat{\theta} \tag{4-31}$$

where ρ is now the mass density function of the combined flow field.

To prepare for the integration over the control surface with constant radius r, we must replace R in equation (4-31) with r, resulting in the velocity at the control surface.

$$V_{cs} = \frac{1}{\rho}\left[\frac{K}{4\pi r^2} - \frac{K(r - a\cos\theta)}{4\pi(r^2 + a^2 - 2ar\cos\theta)^{3/2}}\right]\hat{R} - \frac{1}{\rho}\frac{Ka\sin\theta}{4\pi(r^2 + a^2 - 2ar\cos\theta)^{3/2}}\hat{\theta} \tag{4-32}$$

To determine V_z on the control surface, we can substitute in place of the unit vectors $\hat{R}$ and $\hat{\theta}$ their z-components

$$\hat{R}_z = \cos\theta\,\hat{z} \tag{4-33}$$

$$\hat{\theta}_z = -\sin\theta\,\hat{z} \tag{4-34}$$

This results in

$$V_z = \frac{1}{\rho}\left(\left[\frac{K}{4\pi r^2} - \frac{K(r - a\cos\theta)}{4\pi(r^2 + a^2 - 2ar\cos\theta)^{3/2}}\right]\cos\theta + \frac{Ka\sin\theta}{4\pi(r^2 + a^2 - 2ar\cos\theta)^{3/2}}\sin\theta\right)\hat{z} \tag{4-35}$$

Since I have chosen the control volume to be a sphere of radius r centered on the source, the differential surface area over which we shall integrate is then

$$dA = 2\pi r^2\sin\theta\,d\theta\,\hat{R} \tag{4-36}$$

Calculating the dot product in equation (3-25),

$$(\rho V_{cs}\cdot dA) = 2\pi r^2\sin\theta\left[\frac{K}{4\pi r^2} - \frac{K(r - a\cos\theta)}{4\pi(r^2 + a^2 - 2ar\cos\theta)^{3/2}}\right]d\theta \tag{4-37}$$

Substituting the derived quantities into equation (3-25) we find the force on the control volume in the z-direction to be

$$F_z = 2\pi r^2 \int_0^\pi \frac{\sin\theta}{\rho} \left[\frac{K}{4\pi r^2} - \frac{K(r - a\cos\theta)}{4\pi(r^2 + a^2 - 2ar\cos\theta)^{3/2}} \right]$$

$$\left(\left[\frac{K}{4\pi r^2} - \frac{K(r - a\cos\theta)}{4\pi(r^2 + a^2 - 2ar\cos\theta)^{3/2}} \right] \cos\theta + \frac{Ka\sin\theta}{4\pi(r^2 + a^2 - 2ar\cos\theta)^{3/2}} \sin\theta \right) d\theta \; \hat{z} \qquad (4\text{-}38)$$

This equation can be simplified to result in the following:

$$F_z = \frac{K^2}{8\pi r^2} \int_0^\pi \frac{1}{\rho}$$
$$\left(\left(r^3 - ar^2\cos\theta - \left(r^2 + a^2 - 2ar\cos\theta \right)^{3/2} \right) \right.$$
$$\left. \left(-ar^2 + r^3\cos\theta - \cos\theta \left(r^2 + a^2 - 2ar\cos\theta \right)^{3/2} \right) \sin\theta \, d\theta \right) /$$
$$\left(r^2 + a^2 - 2ar\cos\theta \right)^3$$

$$\hat{z} \qquad (4\text{-}39)$$

Note that the field mass density function, ρ, which is still to be determined, must now remain inside the integral.

Derivation of the Density Field Assuming Light-Speed Everywhere

In the Eulerian fluid continuum theory proposed, the magnitude of the velocity of the steady flow field of edust particles is postulated to be the maximum speed of light *everywhere*. When I choose this, I can then derive the field mass density function. If the magnitude of the velocity equals c everywhere, then from the velocity vector equation (4-31) and the Pythagorean Theorem,

$$c^2 = \left(\frac{1}{\rho} \left[\frac{K}{4\pi R^2} - \frac{K(R - a\cos\theta)}{4\pi(R^2 + a^2 - 2aR\cos\theta)^{3/2}} \right] \right)^2 + \left(\frac{1}{\rho} \frac{Ka\sin\theta}{4\pi(R^2 + a^2 - 2aR\cos\theta)^{3/2}} \right)^2 \qquad (4\text{-}40)$$

Now, moving ρ to the left hand side of the equation, and c to the right side,

$$\rho^2 = \left(\frac{1}{c} \left[\frac{K}{4\pi R^2} - \frac{K(R - a\cos\theta)}{4\pi(R^2 + a^2 - 2aR\cos\theta)^{3/2}} \right] \right)^2 + \left(\frac{1}{c} \frac{Ka\sin\theta}{4\pi(R^2 + a^2 - 2aR\cos\theta)^{3/2}} \right)^2 \qquad (4\text{-}41)$$

Multiplying the component terms, simplifying, and taking the square root, results in the field mass density function

$$\rho = \frac{K}{4\pi c}\left[\frac{1}{R^4} - \frac{2(R - a\cos\theta)}{R^2\left(R^2 + a^2 - 2aR\cos\theta\right)^{3/2}} + \frac{1}{\left(R^2 + a^2 - 2aR\cos\theta\right)^2}\right]^{1/2} \tag{4-42}$$

Now that the mass density of the field is known, it can be plugged into the velocity vector equation (4-31) to determine the velocity vector field. Having equations for the velocity and density fields, we can now plot their characteristic patterns if we assume arbitrary values for the mass flow rate, K, and separation distance, a. Velocity vector, density, and contour plots of the compressible source-sink field are shown in Figure 4-4.

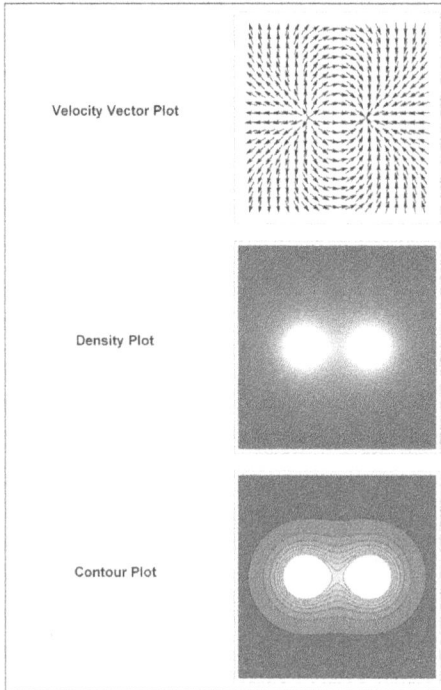

Figure 4-4. Typical velocity vector, density, and contour plots of the compressible, separated source-sink fluid flow field.

Notice that now the velocities associated with this flow are constant in magnitude everywhere, but vary in direction. Furthermore, now the density field shows high concentrations of mass (white representing high density and darker shaded areas representing lower density) at the source and sink. This represents a proper model of reality, i.e. varying concentrations of mass. Also, notice that in the density equation when the value of a = 0, the source and sink centers coalesce, and they annihilate each other as do electrons and positrons, causing the entire field to be without mass.

The Resulting Force Equation for Light-Speed Everywhere

To determine the force on the control volume, we must change R to r in equation (4-42) and substitute it into equation (4-39). This results in

$$
\begin{aligned}
F_z = & \frac{K^2}{8\pi r^2} \int_0^\pi \Big(\big(r^3 - a r^2 \cos\theta - \big(r^2 + a^2 - 2ar\cos\theta\big)^{3/2}\big) \\
& \big(-ar^2 + r^3\cos\theta - \cos\theta \big(r^2 + a^2 - 2ar\cos\theta\big)^{3/2}\big)\sin\theta\, d\theta \Big) \Big/ \\
& \left(\frac{K}{4\pi c}\left[\frac{1}{r^4} - \frac{2(r - a\cos\theta)}{r^2\big(r^2 + a^2 - 2ar\cos\theta\big)^{3/2}} + \frac{1}{\big(r^2 + a^2 - 2ar\cos\theta\big)^2}\right]^{1/2} \right. \\
& \left. \big(r^2 + a^2 - 2ar\cos\theta\big)^3 \right)
\end{aligned}
$$

$$\hat{z} \qquad (4\text{-}43)$$

"Simplifying" this, results in

$$
\begin{aligned}
F_z = & \frac{cK r^2}{2} \int_0^\pi \sin\theta \Bigg(\cos\theta\left(\frac{1}{r^4} - \frac{2(r - a\cos\theta)}{r^2\big(r^2 + a^2 - 2ar\cos\theta\big)^{3/2}} + \frac{(r - a\cos\theta)^2}{\big(r^2 + a^2 - 2ar\cos\theta\big)^3}\right) + \\
& a\left(\frac{1}{r^2\big(r^2 + a^2 - 2ar\cos\theta\big)^{3/2}} - \frac{r - a\cos\theta}{\big(r^2 + a^2 - 2ar\cos\theta\big)^3}\right)\sin^2\theta \Bigg) \Big/ \Bigg[\\
& \frac{1}{r^4} - \frac{2(r - a\cos\theta)}{r^2\big(r^2 + a^2 - 2ar\cos\theta\big)^{3/2}} + \frac{1}{\big(r^2 + a^2 - 2ar\cos\theta\big)^2}\Bigg]^{1/2} \Bigg)\, d\theta\ \hat{z}
\end{aligned}
$$

$$(4\text{-}44)$$

Now that we have the desired integral, the Wolfram *Mathematica* computer program closed form integration function was used in an attempt to solve it. Unfortunately, the program was unsuccessful at solving this complicated integral. However, the change of variables which follows in the next section allowed the integrator to successfully solve the new equation.

Change of Variables

Let there be a change of variables such that

$$u = r^2 + a^2 - 2\,a\,r\cos\theta \qquad\qquad (4\text{-}45)$$

From this, the following useful equations can be derived.

$$\frac{du}{d\theta} = 2\,a\,r\sin\theta \qquad\qquad (4\text{-}46)$$

$$\sin\theta\,d\theta = \frac{du}{2\,a\,r} \qquad\qquad (4\text{-}47)$$

$$\sin^2\theta = 1 - \left(\frac{a^2 + r^2 - u}{2\,a\,r}\right)^2 \qquad\qquad (4\text{-}48)$$

$$\cos\theta = \frac{a^2 + r^2 - u}{2\,a\,r} \qquad\qquad (4\text{-}49)$$

$$r - a\cos\theta = \frac{u - a^2 + r^2}{2\,r} \qquad\qquad (4\text{-}50)$$

Changing the limits of integration we have

$$\text{Lower limit} = r^2 + a^2 - 2\,a\,r \qquad\qquad (4\text{-}51)$$

$$\text{Upper limit} = r^2 + a^2 + 2\,a\,r \qquad\qquad (4\text{-}52)$$

By substituting equations (4-45) through (4-52) into equation (4-44), we have

$$\mathbf{F}_z =$$

$$\frac{cK\,r^2}{2}$$

$$\int_{r^2+a^2-2ar}^{r^2+a^2+2ar} \frac{1}{2ar} \left(\left(\frac{a^2+r^2-u}{2ar} \left(\frac{1}{r^4} - \frac{2\left(\frac{u-a^2+r^2}{2r}\right)}{r^2\,u^{3/2}} + \frac{\left(\frac{u-a^2+r^2}{2r}\right)^2}{u^3} \right) + a \left(\frac{1}{r^2\,u^{3/2}} - \frac{\frac{u-a^2+r^2}{2r}}{u^3} \right) \right.$$

$$\left. \left(1 - \left(\frac{a^2+r^2-u}{2ar} \right)^2 \right) \right) \bigg/ \left[\frac{1}{r^4} - \frac{2\left(\frac{u-a^2+r^2}{2r}\right)}{r^2\,u^{3/2}} + \frac{1}{u^2} \right]^{1/2} du\ \hat{z}$$

(4-53)

This equation can be "simplified" to result in the following:

$$\mathbf{F}_z = \frac{cK}{16\,a^2\,r^4} \int_{r^2+a^2-2ar}^{r^2+a^2+2ar}$$

$$\left(\left(r + u^{1/2} \right) \left(-a^2\,r + r^3 + r\,u - 2\,u^{3/2} \right) \right.$$

$$\left. \left(-a^2 \left(r^2 - r\,u^{1/2} + u \right) + \left(r - u^{1/2} \right)^2 \left(r^2 + r\,u^{1/2} + u \right) \right) du \right) \bigg/$$

$$u^3 \left[\frac{1}{r^4} + \frac{1}{u^2} + \frac{-u+a^2-r^2}{r^2\,u^{3/2}} \right]^{1/2}$$

$$\hat{z}$$
(4-54)

Amazingly, this equation was successfully integrated using the Wolfram Research *Mathematica* closed form integration function, but the solution had a limited range of applicability. Thus, it was also found to be useful to integrate the equation numerically to cover a larger range. The important results follow.

Because of its limited range of applicability, the exact results only closely matched the numerical results when the source and sink were well separated. When they were near each other the results from the closed form solution were complex values, which did not physically make any sense. However, the results of both programs are important. The following plots of the fluid force on the control volume are from the numerical calculations. The importance of the exact solution will become evident later.

As done previously for the incompressible case, the radius of the control volume was chosen to be that of a positron, and the next plots show the fluid force data compared to Coulomb's Law vs. separation distance in different ranges of separation. The plots were normalized in the same manner as the incompressible case.

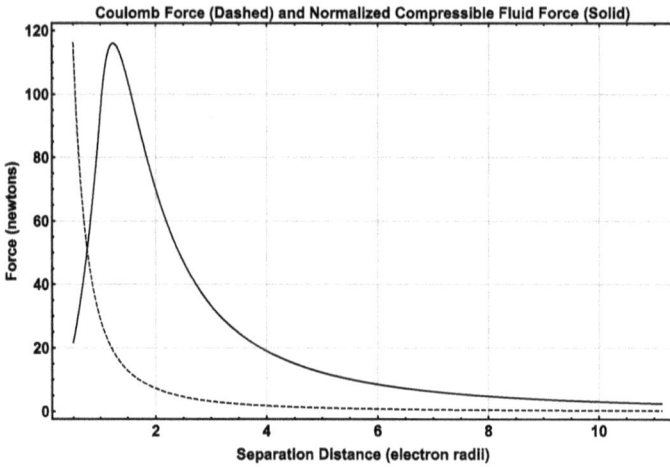

Figure 4-5. The peak-adjusted fluid force F_z on the control volume (Solid) and Coulomb electrostatic force (Dashed) versus values of a/r.

Figure 4-5 shows a peak-adjusted fluid force curve compared to the Coulomb electrostatic force. To initially visualize both curves on the same plot the peak was normalized to the maximum value of the electrostatic force curve by adjusting the mass flow rate, K. Although the two curves look nothing alike, the fluid force on the source's control volume is always positive, pointing toward the sink. This was at least something positive, since the field of the sink caused an attractive force on the source. So far so good. The source and sink attract each other as done by a positron-electron pair. However, unlike the Coulomb force, the fluid force does not go to infinity, but remains finite and has a prominent peak as the separation distance decreases from the maximum separation distance. This is quite different from the previous incompressible fluid flow derivation in Chapter 3 where the fluid force went to infinity. In contrast to the incompressible case, in this compressible fluid flow the force on the control volume is a physically realizable well-behaved phenomenon.

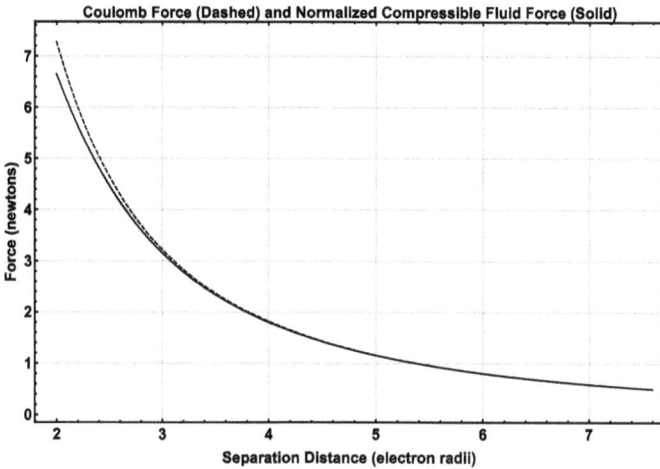

Figure 4-6. The fluid force F_z on the control volume (Solid) normalized to the Coulomb electrostatic force curve (Dashed) at about a/r = 54, versus a/r approximately in the region 2 ≤ a/r < 8.

In Figure 4-6, I have plotted the two curves over a range where the separation distance is about 2 ≤ a/r < 8, and adjusted the mass flow rate so the force curves match on the right side off the graph at a large separation distance which is slightly less than a/r = 54. Note that now both curves match fairly closely, increasing in magnitude with decreasing separation distance, but have measurably different rates of climb when their separation is very small. As opposed to the incompressible case, for this compressible case the magnitude of the fluid force on the control volume is slightly less than the Coulomb force, and increases a little less rapidly than the electrostatic force with decreasing separation distance.

In Figure 4-7, I have plotted the two curves down to a separation distance of a/r = 1/2 and out to the maximum separation distance calculated. In this case, the fluid mass flow rate, K, was adjusted so the fluid force value at the last point on the far right of the graph equaled the electrostatic value. When this was done the value of the mass flow rate, K, was found to be approximately

$$K = 9.69123 \times 10^{-8} \text{ kg/sec.}$$

Notice that now the two curves match fairly well except at the far left where they separate and then diverge from each other. I will discuss this in more detail in the next section after I determine a more precise value of the flow rate, K.

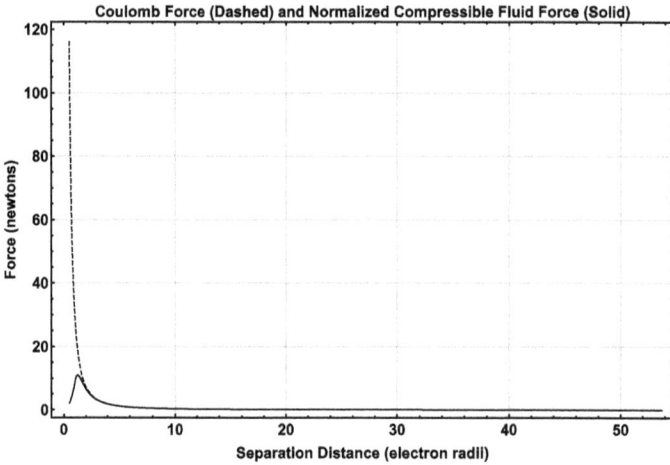

Figure 4-7. The fluid force F_z on the control volume (Solid) normalized to the last point on the Coulomb electrostatic force curve (Dashed) versus a/r.

Rate of Mass Flow, K

As stated earlier, the exact solution is also very important. The exact closed form integration solution was used to attain a more accurate value of the rate of mass flow, K, than found above from the numerical solution. This was achieved by equating the fluid force to the Coulomb force and solving for K at many points of increasing separation distance. As the separation distance was increased, the number of consistent decimal place values in K increased. The calculation was arbitrarily stopped when there were 12 consistent figures in the value of K. This occurred at a separation distance of approximately 5400 electron radii, or roughly 30 percent of the radius of the first Bohr orbit in the hydrogen atom. At that point the value of mass flow rate was

$K = 9.69120697082 \times 10^{-8}$ kg/sec.

This more precise value of mass flow rate was used to plot a close-up of the numerical solution in the region of first few radii of separation as shown in Figure 4-8.

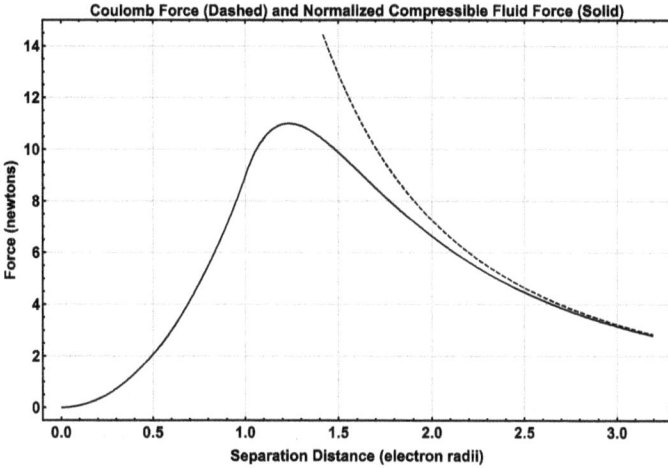

Figure 4-8. The fluid force F_z on the control volume (Solid) and the Coulomb electrostatic force curve (Dashed) versus a/r in the region of the first few radii of separation.

As shown in Figure 4-8, using the new value of K, the magnitude of the fluid force between the source and sink closely approaches the curve of the classical electrostatic Coulomb equation at roughly 3 electron radii, decreasing with approximately the same inverse a-squared decay for larger separation distances. However, as the separation distance decreases toward zero the fluid force on the control volume rises to a peak, remaining finite, then reverses direction downward toward zero. The maximum attractive force of approximately 11.0 newtons occurs at about a/r = 1.235, or 3.48×10^{-15}m of separation.

If this fluid force on the control volume were to truly represent the force between an electron and positron, its characteristics are in stark contrast to the classical electrostatic Coulomb force in two ways: (1) it does not go to infinity, but instead has a maximum force, and (2) the force increases from zero separation distance to a peak, whereas the Coulomb force never increases with separation distance. Note that this unusual phenomenon is reminiscent of the "asymptotic freedom" behavior which has been attributed to a so-called "color force", a force that has never been observed directly [10, p. 685]. The color force is said to act between quarks. It is described as exerting little force at short distances, so that the quarks are like free particles, then increases with more resistance like a rubber band with more separation, and only experiences a strong confining force when they begin to be too far apart. It then vanishes beyond a certain separation distance where the Coulomb force becomes dominant. As shown in Figure 4-8, the force on the compressible source control volume has this behavior. It should be noted here, however, that the control surface of the positron and one which may be assumed around the electron would touch at $a/r = 2$. So, the shapes and densities of the fluid fields of the positron and electron may be assumed to change significantly at this separation distance and as they get closer as the two "particles" merge together. This effect may change the magnitude of the force actually attracting the two point charges. I will discuss the merging of the electron and positron in this range in more detail in Chapter 7.

For now, let's use the value of the calculated mass flow rate, K, to determine the *rest mass* of the electron and positron. The accuracy of this result will justify the effort expended so far by providing confidence in the validity of this theory.

Calculating the Rest Mass of an Electron and Positron

According to the famous physicist, Richard Feynman, the masses of particles used in any scientific theory have not been satisfactorily predicted. They have come from experimental measurements. In his book, *QED: The Strange Theory of Light and Matter*, he states it like this:

> Throughout this entire story there remains one especially unsatisfactory feature: the observed masses of the particles, m. There is no theory that adequately explains these numbers. We use the numbers in all our theories, but we don't understand them --- what they are, or where they

come from. I believe that from a fundamental point of view, this is a very interesting and serious problem. [4, p. 152]

This problem is about to be solved.

In the derivation of the incompressible case in Chapter 3, I needed to specify the density of the flow field, which was chosen to be the average density from the *experimentally* determined value of a point charge's mass. For the case here where the fluid is compressible, I will use the theory to *predict* the mass of the positron (and consequently the electron). Since the theory is based upon the previous postulates and hypotheses, this determination of mass will also be a test of the postulates and hypotheses.

Consider the problem of finding the mass in a spherical control volume of radius r where the density varies as it does in our compressible source and sink. For this problem, we can determine the mass, m, using the density function in equation (4-6), from the integral equation below.

$$m = \iiint \rho \, dV \tag{4-55}$$

$$= \iiint \rho \, R^2 \sin\theta \, dR \, d\phi \, d\theta \tag{4-56}$$

$$= \int_0^\pi \sin\theta \, d\theta \int_0^{2\pi} d\phi \int_0^r \frac{K}{4\pi c R^2} R^2 \, dR \tag{4-57}$$

$$= 2 \cdot 2\pi \cdot \frac{K r}{4\pi c} \tag{4-58}$$

$$m = \frac{K r}{c} \tag{4-59}$$

This simple, elegant result shows that the mass inside the spherical control volume can be determined from the mass flow rate through the source or sink. The mass is the product of the mass flow rate and the radius divided by the speed of light.

Now, the ultimate test of this compressible fluid theory is whether it can produce predictions that agree with experiment. Fortunately, the mass of the electron (and also therefore the positron) has been thoroughly investigated and well established to high precision by many experiments. We are now in a position to test the validity of this theory.

If this theory truly represents the underlying fundamentals of electrostatic phenomena, plugging in the values of the steady mass flow rate,

$$K = 9.69120697082 \times 10^{-8} \text{ kg/sec},$$

the accepted value of the electron's radius [26],

$$r_e = 2.8179403267 \times 10^{-15} \text{ m},$$

and the accepted speed of light [26],

$$c = 2.99792458 \times 10^8 \text{ m/sec},$$

into the simple equation (4-59) should result in the electron's rest mass. We find the rest mass of an electron, m_e, to be

$$m_e = \frac{9.69120697082 \times 10^{-8} \cdot 2.8179403267 \times 10^{-15}}{2.99792458 \times 10^8} \text{ kg}$$

Voilà! The rest mass of the electron derived from this theory is

$$m_e = 9.109382910985 \times 10^{-31} \text{ kg}$$

For comparison, the experimentally accepted value of the electron rest mass [26] is

$$m_e = 9.10938291(40) \times 10^{-31} \text{ kg}$$

The value predicted for the electron's rest mass, derived from fundamental principles in this theory, is correct to *at least all* of the 9 significant figures of the accepted experimentally determined value!

The accuracy of the additional predicted digits for the electron's rest mass from this exact theory appears to be only limited by the precision of the experimentally determined values of the electron's radius and the fundamental physical constants in Coulomb's Law used to determine K. It is no wonder that engineers and physicists stand in awe of the elegance and power of mathematics to describe nature.

Since I have mathematically derived, from fundamental physical principles, the rest mass of the electron and positron which matches the 9 significant figures, the maximum number available, of the accepted experimentally determined value, I conclude that positrons and electrons are not point particles or waves, but behave like and are, in fact, three-dimensional compressible fluid sources and sinks that emit and eliminate edust particles moving at the speed of light in every radial direction. Furthermore, since point charges are entrances and exits that allow mass to enter into and disappear from the universe, they appear to be, or be related to the solutions to the Einstein field equations that have predicted the structure of, so-called "wormholes."

Also, we can now conclude the electron's rest mass is entirely due to its electro-static edust field. There is no fixed "mechanical mass" core with charge distributed throughout its volume or covering its surface, as has been historically conjectured. Furthermore, the force between an electron and positron is not due to the current concept of *two-way* particle exchange which hypothesizes that the electron and positron both emit *and* absorb "force carriers." It is now unambiguously clear that electric fields consist of edust particles, not photons as believed in the so-called standard model. Photons are defined to be wave packets of specific energy and wavelength. The electrostatic fields of electrons and positrons contain no such objects. Neither photons nor waviness plays a part in the cause of the electrostatic force. Furthermore, this result, demonstrating the constant production of edust particles at the source, contradicts the Big Bang theory which says that all mass was created from pure energy at one time about 13.7 billion years ago. Mass appears to be continuously produced in and eliminated from the universe.

Now that we know the value of the rate of mass flow, K, we can plot the magnitude of the fluid density along the centerline between a fixed positron and electron on the z-axis. In the plot in Figure 4-9 the positron is at the origin, and the electron is located at a distance equal to the radius of the first Bohr orbit in a hydrogen atom.

Although the positron is not an adequate replacement for the proton as the nucleus of the hydrogen atom, it does apparently have the same net flow rate attracting the orbiting electron. Given the high values of mass density between the positron and electron as shown in the figure, it is easy to see there is much more than the historically conjectured "empty" space between them.

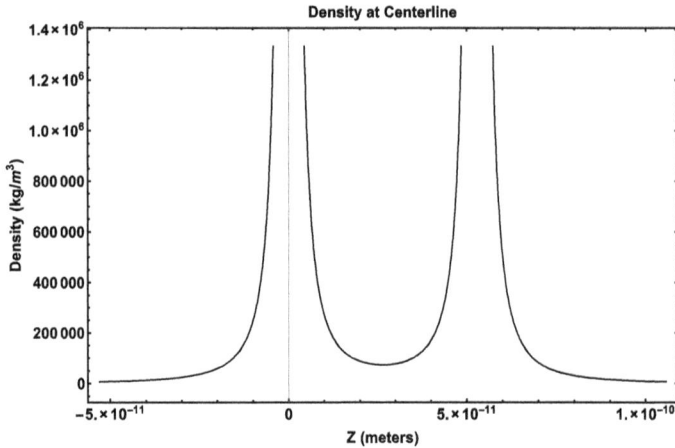

Figure 4-9. The fluid density at the centerline between a positron and electron at a separation distance equal to the radius of the first Bohr orbit.

All There Is?

Since the edust particle appears to be *the* fundamental particle of the universe, let's stop for a moment and consider the enormous number of edust particles that must be emitted and absorbed by all of the charged particles in and around us. If mass emitted from so many electrons or positrons is constantly streaming in the form of edust, then why don't we see it or it block our view? First of all, we don't really "see" what is directly in front of our face. For example, we know that air is normally there, but we can't see it. All we "see" is what our brains picture for us, and that is determined by the physics that takes place within our eyes by the incident photons (composed of edust) that enter them. Beyond that image, we must deduce what is really out there through theory and experimentation. Everything known to us consists of flows of neutral variable density edust currents. Edust particles are streaming in, about and through us everywhere, and, in fact, that's all there appears to be in the reality of our 3-dimensional universe. All of the matter that we see and don't see is edust.

It is a difficult thing to leave one's preconceived understandings and beliefs, and accept such new and radical results. However, the exceptional ability of this theory to calculate the mass of the positron and electron to such accuracy strongly testifies to its validity as a description of nature.

Another much less critical and less definitive way to accept it may be found in the quote below from an Italian writer, Luigi Pirandello:

> Life is full of infinite absurdities, which, strangely enough,
> do not even need to appear plausible, since they are true.

Note: The data used in the calculations above were from 2010 CODATA, and they changed slightly in the 2014 CODATA, shown below:

$e = 1.6021766208(98) \times 10^{-19}$ coulomb

$r_e = 2.8179403227(19) \times 10^{-15}$ m

$m_e = 9.10938356(61) \times 10^{-31}$ kg

The theoretical values below for K and m_e were recalculated using the new data:

$K = 9.6912076733778 \times 10^{-8}$ kg/sec

$m_e = 9.1093835584319 \times 10^{-31}$ kg

The value of the predicted electron's mass does not compare quite as well with the 2014 CODATA, but still has very high accuracy, remaining well within the CODATA error. It seems likely that a value or values used from 2010 CODATA is/are better than the 2014 CODATA.

What is the Macroscopic Electrostatic Force?

It can now be understood that the macroscopic electrostatic force on a point charge is not due to a single force of a field acting specifically on a single elementary particle. It is not a single force at all. It is a perceived single force on a composite particle, which is constantly changing composition, caused by numerous collisions in a field of many edust particles.

Only One Fundamental Force

If edust particles are the only fundamental particles in the universe, the force caused by the collision of two edust particles may be considered a fifth fundamental force to be added to the so-called four "fundamental" forces: electromagnetic, weak nuclear, strong nuclear, and gravitational. However, more properly this is an example of theoretical reduction, where the edust collision force theory reduces the arbitrarily chosen four "fundamental" forces to more basic terms where the electron dust collision force is the *one* fundamental force at the heart of all other forces, and provides a unification of all of the so-called separate "fundamental" forces. Because of this, all phenomena in our universe must be electromagnetic in nature, or if you prefer, purely mechanical.

Chapter 5

Physical Meanings of Electric Charge and Field Strength

What is the Fundamental Electric Charge?

We have found in the source-sink analysis in Chapter 4 that the magnitude of the fluid force between the compressible source and sink closely approximates the classical Coulomb electrostatic force

$$| F_{electrostatic} | = | -\frac{k e^2}{a^2} | \qquad (5\text{-}1)$$

when $a \gtrsim 3\ r$. Now, since there is an inverse a-squared dependence term in front of the integral in the fluid force equation (4-54),

$$F_z = \frac{c K}{16\ a^2\ r^4} \int_{r^2 + a^2 - 2\ a\ r}^{r^2 + a^2 + 2\ a\ r}$$
$$\left((r + u^{1/2})(-a^2\ r + r^3 + r\ u - 2\ u^{3/2}) \right.$$
$$\left. (-a^2(r^2 - r\ u^{1/2} + u) + (r - u^{1/2})^2(r^2 + r\ u^{1/2} + u))\ du \right) /$$
$$u^3 \left[\frac{1}{r^4} + \frac{1}{u^2} + \frac{-u + a^2 - r^2}{r^2\ u^{3/2}} \right]^{1/2}$$
$$\hat{z} \qquad (4\text{-}54)$$

the integral itself must be approximately a constant for any given value of $a \gtrsim 3\ r$. Therefore, in this region the magnitude of the fluid force can be put into the form

$$F_z = C \frac{K}{a^2} \qquad (5\text{-}2)$$

where C is a constant.

Equating the fluid and Coulomb forces, we have

$$\frac{k\,e^2}{a^2} = C\frac{K}{a^2} \tag{5-3}$$

Now, solving for the value of the constant, C, we find that

$$C = \frac{k\,e^2}{K} \tag{5-4}$$

and solving equation (5-4) for the elementary charge, e, we attain that

$$e = \sqrt{\frac{CK}{k}} = \sqrt{4\,\pi\,\epsilon_0\,CK} \tag{5-5}$$

This equation shows that the fundamental property of electrons and positrons called *elementary charge* is proportional to the square root of the mass flow rate, K. Now, we know what charge is and what it is not. "Charge" is *neither*, as suggested in many textbooks, a quantity of some mystical substance that is confined to a very small but non-zero volume, *nor* can it exist on or be spread over a surface like peanut butter. *Elementary charge, e, is simply a measure of the fundamental mass flow rate, K.*

It has been said that the mass of an electron is very small compared to its "charge." Now, we can show a more apples-to-apples comparison. The mass flow rate, K, emanating from a source point charge has been shown to be $9.69120697082 \times 10^{-8}$ kilograms/second. If we divide the electron's rest mass m_e into the mass flow rate K, the result shows that the equivalent mass of 1.06387×10^{23} electrons flows through *one* electron's control volume every second. The mass of an electron *is indeed* extremely small compared to the rate of mass flow through it.

In the case of a body containing many point charges, the net "charge" is simply a measure of the net mass flow rate from that body. Note, however, a net mass flow rate neither increases nor decreases the body's mass, because mass is *not* locally conserved. Edust particles are continually created and destroyed, or otherwise produced into and transported out of our currently known 3-D reality. This phenomenon creates a physical explanation for the behavior in particle reactions where it is said by M. J. G. Veltman in his book, *Facts and Mysteries in Elementary*

Particle Physics, that:

> All charged particles can emit or absorb photons, but they remain the same particle, for example an electron may become an electron and a photon. [27, p. 53]

From the description of point charges above, they should also produce a wavy edust field if oscillated. This wavy edust field may be considered a photon emitted by a single charged particle, similar to the wave packets formed when electrons change orbits in atoms. However, as will be described later in the next chapter, these latter wave-packet photons appear to be disturbances in the edust flow field caused by the combined motions of two charged particles, the electron *and* nucleus of atoms.

What is Electric Field Strength?

Since we know from many elementary textbooks, and equation (2-1), that the magnitude of the electric field strength, E, of a single positive point charge is

$$E = \frac{ke}{R^2} \qquad (2\text{-}1)$$

we can substitute the value of e from equation (5-5) to find that,

$$E = \frac{\sqrt{CkK}}{R^2} \qquad (5\text{-}6)$$

and show the value of the electric field strength in terms of a fluid flow composed of edust particles moving at the speed of light.

Since we know the mass density function, ρ, of the compressible source is

$$\rho = \frac{K}{4\pi c R^2} \qquad (4\text{-}6)$$

it follows from eliminating R^2 from the last two equations that we can obtain the relation between the mass density function and electric field strength,

$$\rho = \frac{\sqrt{K/Ck}}{4\pi c} E \tag{5-7}$$

or,

$$E = \frac{4\pi c}{\sqrt{K/Ck}} \rho \tag{5-8}$$

Therefore, electric field strength *is* proportional to the mass density of the edust field.

The analysis above gives physical meanings to the concepts of electric charge and electric field strength in terms of the flow of particles of mass moving at the speed of light. The equations (5-5) and (5-8) which link electric charge and electric field strength to mass flow rate and mass density, respectively, are two triumphs of relating previously believed disparate quantities — one of the great aims of physics.

The fact that an electric field is a flow of fluid has been consistently misunderstood and rejected in textbooks. A common example of this can be found in a book by H. M. Schey, while discussing Gauss' law:

> The electric field 'flows' out of a surface enclosing charge, and the 'amount' of this 'flow' is proportional to the net charge enclosed. Warning: This is not to be taken literally; the electric field is not flowing in the sense in which fluid flows. It is merely picturesque language intended to aid us in understanding the physics in Gauss' law. [28, p. 33]

Many textbooks shall need revision.

Chapter 6

On Quantum Mechanics

Motivation

In Chapter 4, I showed that the electron has a definite structure and the characteristics of a particle and a field, both of which are composed of flowing particles of mass, edust particles. Also, the electron does not have any intrinsic wave characteristics. This disagrees with the wave-particle duality nature of particles central to quantum mechanics. Quantum mechanics essentially throws out any picture of a particle with any firm description, and replaces it with a wave that is spread out in space without a specific location. The wave is called a "matter" wave by some, described by a wave function, which can be used to determine the probability of finding the electron, and that is said to be all that exists until a measurement is made. When a measurement is made the wave function is said to "collapse," and the electron is then said to have a known location. One description of this is from a website discussing it, and credited to John D. Norton:

> In the strange realm of electrons, photons and the other fundamental particles, quantum mechanics is law. Particles don't behave like tiny balls, but rather like waves that are spread over a large area. Each particle is described by a "wavefunction," or probability distribution, which tells what its location, velocity, and other properties are more likely to be, but not what those properties are. The particle actually has a range of values for all the properties, until you experimentally measure one of them — its location, for example — at which point the particle's wavefunction "collapses" and it adopts just one location.
>
> But how and why does measuring a particle make its wavefunction collapse, producing the concrete reality that we perceive to exist? The

issue, known as the measurement problem, may seem esoteric, but our understanding of what reality is, or if it exists at all, hinges upon the answer. [29]

Some of the instructors of quantum mechanics have told their students to not even question this, just learn how to do the mathematical calculations with the wave function to determine probabilities, and accept it. This thought is generally stated at the beginning of most every quantum mechanics book. Below is one of those statements by Richard Feynman:

> ... I think I can safely say that nobody understands quantum mechanics. ... Do not keep saying to yourself, if you can possibly avoid it, 'But how can it be like that?' because you will get 'down the drain,' into a blind alley from which nobody has yet escaped. Nobody knows how it can be like that. [30]

Does a particle exist without being measured, or does a measurement create a particle and determine its location? I have already explained why point charges are different than previously perceived, which might have caused this question to begin with, but I shall completely answer this question, and escape from the "blind alley" of quantum mechanics in this chapter. However, first I must lay some ground work by discussing the empirical foundation for testing atomic models, and create a new model of atoms.

The Empirical Foundation of the Atomic Electromagnetic Spectrum

One of the primary tests of the validity of atomic models is based on the model's ability to predict the experimentally observed electromagnetic spectrum an atom produces due to its electrons transitioning between atomic energy levels. In the late 1800s the regularity of the hydrogen spectrum, determined by atomic spectroscopy, led several researchers to look for an empirical formula to represent the wavelengths of its spectral lines. Several partial formulas were found, and those discoveries led to a search for more complete empirical formulas to describe the spectra of other elements. Most of this work on many chemical elements was done by the Swedish physicist, Johannes Rydberg, who presented a formula from his work in 1888. The following is the Rydberg equation for the hydrogen atom,

$$\frac{1}{\lambda} = R \left(\frac{1}{n_l^2} - \frac{1}{n_u^2} \right)$$
(6-1)

where λ is the photon wavelength, R is the Rydberg constant, and n_l and n_u are the principal quantum numbers of the lower and upper orbits, which are positive integers with $n_u > n_l$. All of the spectral lines of the hydrogen atom can be computed very accurately from this one general formula [31, p. 181].

From the Rydberg equation, in 1908 Walter Ritz developed what is known as the Ritz Combination Law that essentially says that the photon frequency emitted from an atom during a large electron jump, when an electron jumps from one orbit over one or more other orbits before settling in a final orbit, can be determined from the sum of photon frequencies that would have been emitted if the electron had made stops at the orbits in between.

For example, if we put equation (6-1) in the form of a photon frequency, v, where λ = c/v, we have

$$v = cR \left(\frac{1}{n_l^2} - \frac{1}{n_u^2} \right)$$
(6-2)

Suppose now that an electron jumps from an arbitrary higher upper orbit 3 to a lower orbit 1, and skips an intermediate orbit 2. Then, the emitted photon frequency is

$$v_{31} = cR \left(\frac{1}{n_1^2} - \frac{1}{n_3^2} \right)$$
(6-3)

But if the electron had made a stop at an intermediate orbit, the two photons released would have had frequencies

$$v_{32} = cR \left(\frac{1}{n_2^2} - \frac{1}{n_3^2} \right)$$
(6-4)

$$v_{21} = cR \left(\frac{1}{n_1{}^2} - \frac{1}{n_2{}^2} \right) \tag{6-5}$$

And, it is easy to see that

$$v_{31} = v_{32} + v_{21} \tag{6-6}$$

This law is said to still be in use today.

Although the values of photon frequencies can be attained from the Rydberg equation and Ritz Combination Law, the reason each photon frequency is equal to the difference between the two terms, each of which may be considered to be some type of *abstract characteristic frequency* of an orbit or atomic state, has not been understood. No classical theory of radiation from orbiting electrons has been able to account for these frequencies. Paul Dirac stated this was "quite unintelligible from the classical standpoint" in his book, *Principles of Quantum Mechanics*:

> ... if an atomic system has its equilibrium disturbed in any way and is then left alone, it will be set in oscillation and the oscillations will get impressed on the surrounding electromagnetic field, so that their frequencies may be observed with a spectroscope. Now whatever the laws of force governing the equilibrium, one would expect to be able to include the various frequencies in a scheme comprising certain fundamental frequencies and their harmonics. This is not observed to be the case. Instead, there is observed a new and unexpected connexion between the frequencies, called Ritz's Combination Law of Spectroscopy, according to which all the frequencies can be expressed as differences between certain terms, the number of terms being much less than the number of frequencies. This law is quite unintelligible from the classical standpoint. [32, p. 2]

The inability to determine photon frequencies from a classical standpoint was one of the problems that led to the development of quantum mechanics. To explain why the Rydberg and Ritz equations work, and determine photon frequencies from a *classical* standpoint, I shall continue with a short background on atomic models.

Short Historical Background on Atomic Models

Several theoretical atomic models were developed shortly after the beginning of the 20th century which could be tested by the Rydberg equation. One of the well known models was proposed in 1904 by J.J. Thomson, and another was proposed in 1911 by Ernest Rutherford. J.J. Thomson proposed the so-called "plum pudding" model of the atom in which electrons were thought to be evenly distributed throughout a sphere of positive charge. In their lowest energy state they were fixed at equilibrium positions, and were thought to vibrate about those positions when excited. However, agreement with the experimentally observed spectra was lacking. Rutherford developed a model of the atom in which a very small nucleus was orbited by electrons, much like the orbits of planets about the sun. In this arrangement, however, the electrons could be in any orbit, and produce any continuous frequency spectrum compatible with Newton's laws of motion. This also is not in agreement with the discrete frequency spectrum known to be emitted by atoms. In 1913, Niels Bohr developed a theory of the hydrogen atom and an equation that were in accurate quantitative agreement with the Rydberg formula.

Bohr's Theoretical Derivation of the Rydberg Formula

The content of this section can be found in many textbooks. Nothing in it is particularly new. However, this section will provide the basis to develop a new form of Bohr's theoretical Rydberg equation, an enlightening interpretation of it, and extend the Bohr theory to create a new concept of atoms. Since the anticipated audience for this book should be very familiar with the development of the Bohr theory and Rydberg formula, I will make it brief, and only provide the necessary parts.

Two of Bohr's postulates state that an atom can only gain or lose energy by electrons jumping from one allowed orbit to another, and that they absorb or emit electromagnetic radiation in the form of a photon with a frequency determined by the energy difference between the allowed orbital levels. The following equation, known as Bohr's relation, equates the difference between atomic energy levels to the photon frequency times Planck's constant, h.

$$E_u - E_l = h\nu \tag{6-7}$$

or

$$\nu = \frac{E_u - E_l}{h} \tag{6-8}$$

where E_u is the energy of the atom with the electron in an upper orbit, E_l is the energy of the atom with the electron in a lower orbit, and ν is the frequency of the energy lost or gained due to electromagnetic radiation. This is essentially in agreement with Einstein's postulate that the frequency of a photon of electromagnetic radiation is equal to the energy of the photon divided by Planck's constant, but it adds how the atom loses or gains its energy. Bohr attributed the energy lost or gained by the atom to the electron changing orbits.

Other assumptions by Bohr were: the electron moves at non-relativistic speeds; the proton mass is essentially at rest; the electron can only move in specific classical circular orbits about the nucleus; and the atom does not radiate energy while in those specific orbits. Thus, the energy of the atom would remain constant while in a stationary orbit. He also determined that the energy spacing between orbital levels requires that the angular momentum, L, be restricted to only integer multiples of Planck's constant divided by 2 times pi,

$$L = n\frac{h}{2\pi} = n\hbar \tag{6-9}$$

where $n = 1, 2, 3, \ldots$ are the principal quantum numbers. Given that an electron with mass m and speed v in a circular orbit of radius r has angular momentum

$$L = mvr \tag{6-10}$$

Bohr derived the equation,

$$mvr = n\hbar \tag{6-11}$$

Now, applying Newton's force equation, $F = ma$, to an electron in circular orbit subject to the electromagnetic force of a proton, Bohr derived the equation

$$ke^2/r^2 = mv^2/r \qquad (6\text{-}12)$$

These last two equations were solved for the two unknowns, radius and velocity, to attain the allowed orbital radii, r_n, and electron velocities, v_n, of Bohr's theory.

$$r_n = \frac{n^2 \hbar^2}{k\,m\,e^2} \qquad (6\text{-}13)$$

$$v_n = \frac{k\,e^2}{n\,\hbar} \qquad (6\text{-}14)$$

For this system the total mechanical energy, E, excluding the rest energies of the electron and proton, is the sum of the kinetic energy, K.E., and the electrostatic potential energy, P.E., or

$$E = K.E. + P.E. \qquad (6\text{-}15)$$

Since the proton was assumed to be essentially at rest, this equation becomes

$$E = mv^2/2 + (-\,ke^2/r) \qquad (6\text{-}16)$$

Substituting the allowed orbital radii, r_n, and electron velocities, v_n, into this equation, Bohr found the only possible energies of the hydrogen atom to be

$$E_n = -\,\frac{k^2\,e^4\,m}{2\,n^2\,\hbar^2} \qquad (6\text{-}17)$$

Now from Bohr's relation

$$E_u - E_l = h\nu \qquad (6\text{-}7)$$

the Bohr theoretical prediction of the frequencies of photons released when electrons jump from upper to lower orbits can be determined. That is,

$$v = \frac{1}{h}\left(-\frac{k^2\,e^4\,m}{2\,n_u^2\,\hbar^2} + \frac{k^2\,e^4\,m}{2\,n_l^2\,\hbar^2}\right) \tag{6-18}$$

Now, by rearranging the terms and noting that $\lambda = c/v$, the Bohr equation can be put into the same form as the Rydberg equation,

$$\frac{1}{\lambda} = \frac{k^2\,e^4\,m}{4\,\pi\,\hbar^3\,c}\left(\frac{1}{n_l^2} - \frac{1}{n_u^2}\right) \tag{6-19}$$

where the Bohr theoretical Rydberg constant turns out to be

$$R = \frac{k^2\,e^4\,m}{4\,\pi\,\hbar^3\,c} \tag{6-20}$$

Although this constant is in good agreement with the empirical Rydberg constant, it can be improved by accounting for the fact that the proton mass is not infinite. The exact equation of motion of the Bohr hydrogen electron about the center of mass can be determined by replacing the electron mass with an effective mass called the *reduced mass*, μ. In fact, all the relations of the Bohr theory may be retained if the electron mass, m, is replaced by the reduced mass,

$$\mu = \frac{m_e\,m_p}{m_e + m_p} \tag{6-21}$$

where m_e is the mass of the electron, and m_p is the mass of the proton.

The theoretical derivation of the Rydberg formula for the spectral lines of the hydrogen atom is the Bohr model's key success. The Bohr model provided a theoretical basis in terms of fundamental physical constants for the empirical Rydberg formula, and the value of the theoretical Rydberg constant matches very closely the value determined experimentally [31, p.188-189].

Photon Frequencies - "Unintelligible From the Classical Standpoint"?

Although the Bohr theory was very successful, it did not answer all of the questions about the hydrogen atom. In fact, it appeared to conflict with classical thinking in several areas. One problem is that although the values of the photon frequencies can be attained from the Bohr theoretical Rydberg equation, the reason they are proportional to the difference between the energies of two different orbits, or atomic states, has not been explained in any classical sense.

Another issue concerns the principle of causality. Apparently, even before Bohr was able to have his paper on the hydrogen atom published, Rutherford raised the issue that the Bohr relation has a problem with the principle of causality. According to Bohr's theory, when an electron jumps from an upper stationary orbit to a lower stationary orbit, the electron emits a photon with frequency v during the transition according to equation (6-8),

$$v = \frac{E_u - E_l}{h} \qquad (6\text{-}8)$$

Rutherford felt that it was necessary for the electron to know before it settled in its final stationary orbit in which orbit it was going to stop to emit the correct frequency [33, p. 212]. Certainly, this seems to be a logical conclusion. Even to date, quantum mechanics has been unable to satisfactorily explain in any classical sense the relationship between the Newtonian orbital frequencies and photon frequencies, or their relationship to the electron's jumps to different orbits. Consequently, quantum mechanics' answer to this problem is to simply assert that Newton's laws of motion are not applicable in atoms, and causality does not exist there.

R. Feynman discusses this problem in his book, *QED: The Strange Theory of Light and Matter*, and concludes, as does quantum mechanics, that Newton's laws of motion were wrong in atoms:

> Attempts to understand the motion of the electrons going around the nucleus by using mechanical laws --- analogous to the way Newton used the laws of motion to figure out how the earth went around the sun

--- were a real failure: all kinds of predictions came out wrong. ... Working out another system to replace Newton's laws took a long time because phenomena at the atomic level were quite strange. One had to lose one's common sense in order to perceive what was happening at the atomic level. Finally, in 1926, an "uncommon-sensy" theory was developed to explain the "new type of behavior" of electrons in matter. It looked cockeyed, but in reality it was not: it was called the theory of quantum mechanics. [4, p. 5]

In the following sections in this chapter, I shall: derive a new atomic model by extending the Bohr theory; determine the cause for the creation of photons and photon frequencies using classical Newtonian mechanics; provide further support to re-establish causality in the microscopic world; and explain quantum mechanical phenomena in a common classical sense.

A New Form of the Bohr Theoretical Rydberg Equation

In an effort to bring more reality to the probabilistic results of quantum mechanics theory, I will begin by deriving a new form of Bohr's theoretical Rydberg equation.

In quantum theory, it is said that an electron cannot be determined to follow a definite orbit. However, for every possible orbit, i.e. every "n", in the Bohr model, the radii and frequencies of those orbits are identical to the classical frequencies derived from Newton's laws of motion. Thus, all of the orbits in the Bohr model are members of the possible orbits predicted classically. What has been disturbing is that when electrons change orbits from an upper orbit to a lower orbit the frequencies of the photons which propagate away have not been derivable from the frequencies of the upper and lower orbits. I will now develop an extension to Bohr's theory, and show that all of the photon frequencies can be made to agree with classical physics thinking.

If we substitute $\lambda = c/v$ into equation (6-19), we can put the Bohr theoretical Rydberg equation in the form of photon frequencies instead of wavelengths,

$$v = \frac{k^2 e^4 m}{4 \pi \hbar^3} \frac{1}{n_l^2} - \frac{k^2 e^4 m}{4 \pi \hbar^3} \frac{1}{n_u^2} \tag{6-22}$$

and now the two terms on the right-hand side of this equation can be considered to be some type of *abstract characteristic frequency* of the lower and upper orbits. Because this equation contains the difference between the abstract characteristic frequencies, it hints that the photon frequency could possibly be a beat frequency related to these two frequencies. Unfortunately, the two frequencies are *not* the orbital frequencies.

Being certain that the frequencies of photons must be related directly in some *classical* manner to the upper and lower *orbital* frequencies, I decided to study them more closely. The first four Bohr orbital and abstract characteristic frequencies are shown in the graph of Figure 6-1.

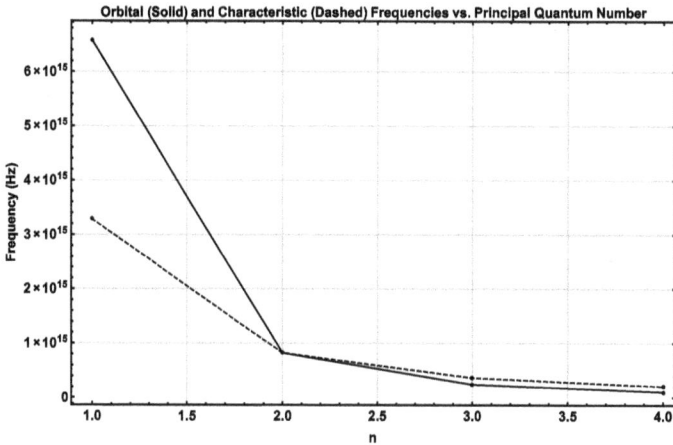

Figure 6-1. Orbital and Characteristic Frequencies for the first four Bohr orbits vs. Principal Quantum Number, *n*.

It is interesting that the two lines in Figure 6-1 connecting the discrete orbital and characteristic frequency values have the same value and cross at $n = 2$. To be more specific, the numerical values of the frequencies plotted in Figure 6-1 are shown in Table 6-1.

By comparing the frequency values, it can easily be determined that if we multiply the Bohr orbital frequencies by the factor $n/2$, the resulting values are *exactly* the abstract characteristic frequencies!

n	1	2	3	4
Bohr Orbital	6.5761×10^{15}	8.22013×10^{14}	2.43559×10^{14}	1.02752×10^{14}
Characteristic	3.28805×10^{15}	8.22013×10^{14}	3.65339×10^{14}	2.05503×10^{14}

Table 6-1. Orbital and Characteristic Frequency values (Hz) and their Principal Quantum Numbers for the first four Bohr orbits.

It turns out that this new finding can be derived directly from the Bohr theory, making the factor $n/2$ applicable to *all* of the Bohr orbital and corresponding characteristic frequencies. I shall derive that now.

Using Newton's force law, the equation for any orbital frequency, f, of the electron in circular motion around a proton turns out to be

$$f = (e/2\pi)\sqrt{k/mr^3} \qquad (6\text{-}23)$$

However, the Bohr theory only allows the specific orbital radii given by equation (6-13) repeated below,

$$r_n = \frac{n^2\hbar^2}{k\,m\,e^2} \qquad (6\text{-}13)$$

Thus, the allowed orbital frequencies are designated by the principal quantum number, n, by the equation

$$f_n = (e/2\pi)\sqrt{k/mr_n^3} \qquad (6\text{-}24)$$

Now, if we insert the allowed orbital radii, r_n, into the frequency equation for f_n, and re-arrange some of the variables, in particular by having a numerical constant and a portion of the n-factor on each side of the equation, we can write it as

$$\frac{n\,f_n}{2} = \frac{k^2\,e^4\,m}{4\,\pi\,\hbar^3}\,\frac{1}{n^2} \tag{6-25}$$

Now, if we compare the right-hand side of equation (6-25) to the terms on the right-hand side of equation (6-22) repeated below,

$$\nu = \frac{k^2\,e^4\,m}{4\,\pi\,\hbar^3}\,\frac{1}{n_l^2} - \frac{k^2\,e^4\,m}{4\,\pi\,\hbar^3}\,\frac{1}{n_u^2} \tag{6-22}$$

we can immediately put the Bohr theoretical Rydberg equation into the following *new* form,

$$\nu = \frac{n_l\,f_l}{2} - \frac{n_u\,f_u}{2} \tag{6-26}$$

where the frequency of a photon, ν, can be determined by simply knowing the principal quantum numbers and the *orbital* frequencies of the upper and lower orbits. This new result which relates the photon frequency to the orbital frequencies of the upper and lower orbits is encouraging, but now I need to make some further sense of it because of the $n/2$ factors to determine what it really means physically.

An Oscillatory Atomic Model

As stated at the beginning of this chapter, one of the primary tests of the validity of atomic models is based on the model's ability to predict the experimentally observed electromagnetic spectrum the atom produces. The Bohr model was successful in doing this, but was not successful in explaining in any classical sense why. I shall now develop a *new* atomic model that *is* capable of explaining in classical terms why this is possible.

Now, in many modern derivations of the Bohr theory an equation developed by

Louis de Broglie, and known as the *de Broglie relation* is used instead of Bohr's equation that restricts angular momentum to certain values. The following equation is the *de Broglie relation* [31, p. 130],

$$\lambda = h/p = h/mv \qquad\qquad (6\text{-}27)$$

where p is the momentum of macroscopic particles. Although the de Broglie relation was derived years after the Bohr theory, and Bohr's angular momentum requirement and the de Broglie relation are completely equivalent, the de Broglie relation is commonly used now to derive the theory because it has an experimental foundation from diffraction experiments. The de Broglie relation introduces the so-called wave nature of matter used in the current theory of quantum mechanics:

> ... de Broglie ... said waves are particles and particles are waves. ... This wave duality remains central in quantum physics. [2, p.184]

Also, according to de Broglie all particles of mass have an accompanying wave [34]. When the wave aspect of the electron, governed by the de Broglie relation, is incorporated into the Bohr theory, the electron's wave is mathematically described as a standing wave that must have an integral number of whole wavelengths fitted around the circular orbital circumference. Furthermore, the acceptable orbits, the Bohr orbits, begin with one wavelength per orbit and incrementally increase the number of wavelengths per orbit by one. A representation of one of de Broglie's stationary electron waves of six wavelengths is shown in Figure 6-2.

Since de Broglie hypothesized that a wave must accompany the electron, there has been much speculation about the nature of the wave. Some authors have said that the electron "drags" the wave along, while others say the wave is just somewhere in the "neighborhood" of the electron. Other authors have interpreted the wave as some sort of a resonance. Some authors consistently refer to the wave as a matter wave, as if the mass of the electron is somehow spread around the circumference of the orbit, and the electron only exists as a particle if observed in a measurement.

Now, a standing wave is normally shown to be formed by two waves moving in opposite directions with constructive interference. How this occurs for particles of mass has not been explained. Some authors have asked the question: What actually waves? This also has not been explained. With regard to the wave function in quantum mechanics, this question has even been said to be inappropriate to ask since the wave is characterized as just a complex mathematical function, used to determine the probability of finding a particle. However, something physical must provide a wave to exhibit the constructive and destructive phenomena in diffraction experiments.

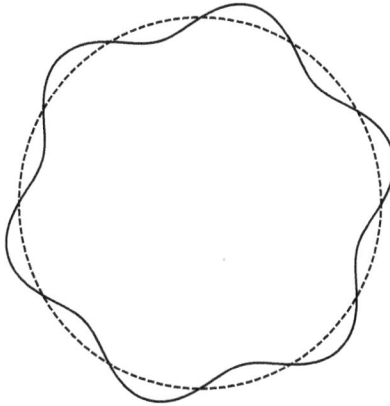

6 λs fitted to an orbit of length $2\pi r_6$

Figure 6-2. Example stationary electron wave of six wavelengths.

Instead of a standing wave, in some texts on quantum mechanics the wave function is characterized as a traveling wave [10, p. 178]. The classical type of this kind of wave is generally explained using an example of a taut string being oscillated by a sinusoidal forcing function at one end, causing energy to propagate down the string. But we know that an electron is not a string, and furthermore at its non-relativistic orbital velocities, it is practically spherical.

Also, from Chapter 4, we know that electrons are neither strictly particles, nor intrinsically have wave properties. That is, if an electron were stationary or were to move with a constant velocity, we would not expect it to have an accompanying wave. This does not agree well with de Broglie's hypothesis. However, electrons do have edust fields, and may produce waves in their fields, if they are oscillated. Also, de Broglie's relation concerning particles exhibiting wave-like characteristics turned out to work very well for electrons in atoms. So, what causes quantum wave mechanics, which is heavily based on de Broglie's relation, to work so well? Suppose I make the following hypothesis:

Hypothesis VI:

The reason that wave mechanics for electrons works so well is <u>not</u> that electrons have intrinsic wave-like properties or have accompanying waves, but that electrons are forced to oscillate radially when they orbit a nucleus or pass by nucleons.

This would provide the answer to the question: What is waving?

If we think back about the de Broglie orbital wave pattern in Figure 6-2, it is a wave pattern which is *spatially* dependent, not a temporally dependent wave. That is, its wavelength and spatial frequency depends on the length of the orbit, the circumference, not on a time period of oscillation. However, the electron is moving along the circumference with a specific temporal orbital frequency according to Newton's laws. Now, if an electron is oscillating while in orbit in the hydrogen atom, de Broglie's requirement that there are n complete cycles in each orbit would require the electron to oscillate radially n times in each nth-orbit. If the electron is both orbiting *and* oscillating, the resulting temporal frequency of electron oscillation would be the spatial frequency, n-times per orbit, n_{orbit}, multiplied by the orbital frequency of that orbit, orbits per unit time, f_{orbit}, resulting in the temporal oscillation frequency of $n_{orbit}f_{orbit}$. This product is, in fact, precisely what is found in the new form of the theoretical Rydberg equation, (6-26), excluding the factor $1/2$. Thus, from this reasoning, I may conclude that the two abstract characteristic frequencies in the new form of the Bohr theoretical Rydberg equation are each one-half of the electron *radial* oscillatory frequencies in their orbits.

Now, the results of the next major improvement to quantum mechanics after the Bohr theory, i.e. Schrödinger's equation, support this hypothesis. They show that the hydrogen atom has exactly the same discrete values of possible energies as did

Bohr, but instead of strictly circular orbits there is a wide region of significant probability of finding the electron at a range of radial locations around the various Bohr orbital radii, rather than just staying at those specific radii predicted by the Bohr theory [10, p. 245]. Furthermore, there is overwhelming evidence that shows the quantum mechanical prediction for the angular momentum of the electron in the first Bohr orbit is zero! [10, p. 254] The proposed extension to the Bohr theory, that the electron oscillates about its orbits, helps to explain all of these results. In fact, it implies the motion in the first Bohr orbit is entirely radial in that state. An entirely radial oscillation for this case, which could take place along any direction in space, would correspond to a spherically symmetric probability density which is predicted by quantum mechanics and observed experimentally [10, p. 254].

Theory of the Frequency of Photons

Now, assuming the last hypothesis is true, according to the new form of the Bohr theoretical Rydberg equation the frequency of a photon is the difference between the oscillatory frequencies of the electron in the lower and upper orbits times 1/2. This mathematical form for a frequency is *exactly* like that found in a beat frequency phenomenon. This phenomenon is caused when two sinusoidally oscillating waves with different frequencies are added together. The two waves will form a new wave with an amplitude which oscillates periodically with a frequency that is one-half the difference between the two frequencies. This is derived in many textbooks concerning beat frequency, e.g. [35, p. 964], as shown below.

Consider the superposition of two waves of equal displacement amplitude and initially no phase difference (for convenience and definiteness), but of different frequencies.

$$y_1 = A \sin 2\pi f_1 t \qquad\qquad (6\text{-}28)$$

$$y_2 = A \sin 2\pi f_2 t \qquad\qquad (6\text{-}29)$$

The resultant displacement is

$$y = y_1 + y_2 = A(\sin 2\pi f_1 t + \sin 2\pi f_2 t) \tag{6-30}$$

Using the trigonometric identity

$$\sin a + \sin b = 2 \cos \left(\frac{a-b}{2}\right) \sin \left(\frac{a+b}{2}\right) \tag{6-31}$$

equation (6-30) can be put in the form

$$y = (2A\cos 2\pi(\tfrac{f_1 - f_2}{2})t) \sin 2\pi(\tfrac{f_1 + f_2}{2})t \tag{6-32}$$

or

$$y = B \sin 2\pi f t \tag{6-33}$$

The displacement can now be interpreted as the product of two sinusoidal functions, one which has an "amplitude", B, with a magnitude of twice the magnitude of the original waves oscillating at the frequency $(f_1 - f_2)/2$, and $\sin(2\pi f t)$ which oscillates at the higher frequency, $f = (f_1 + f_2)/2$. Thus, the envelope of the composite wave oscillates at the frequency $(f_1 - f_2)/2$. If the two different frequencies of the two original displacement waves have the lower orbit and upper orbit electron oscillatory frequencies, the amplitude envelope frequency would be exactly the frequency of the emitted photon!

From this analysis, I conclude that the two abstract characteristic frequencies, each of which includes a factor of 1/2 and has the dimension of a frequency, are *not* physical frequencies at all. The appropriate representation and interpretation of the photon frequency is the single combined term, i.e. an oscillatory frequency in the lower orbit minus an oscillatory frequency in the upper orbit divided by 2.

$$v = \frac{n_l\, f_l - n_u\, f_u}{2} \tag{6-34}$$

The photon frequency is the real physical frequency resulting from the combination of two real physical oscillatory frequencies in a beat phenomenon, *not* a result of the simple difference between two physical frequencies.

Now the question becomes: How is a photon formed from two different oscillatory frequencies, an upper orbit oscillatory frequency and a lower orbit oscillatory frequency, at the *same time* to cause this beat phenomenon?

Theory of the Photon-Creation Process

Neither the old quantum theories (Bohr and Sommerfeld) of the atomic structure, nor the newer quantum theories (Edwin Schrödinger, Paul Dirac, and quantum electrodynamics) give any *physical* details of the electron's quantum jump to a lower stationary state, or of the photon-creation process.

Instead, the new quantum theory adds confusion to the jump by assuming that a single particle can be in two different states of motion at the same time, as related below in Kenneth Ford's book, *The Quantum World*:

> ... quantum mechanics ... permits particles to be in two or more states of motion at the same time; it allows a particle to interfere with itself; ... [2, p. 247]

Several well respected textbooks present examples (e.g. [10, p. 166-167] and [15, p. 29]) describing how a particle, i.e. a single electron, could potentially emit a photon where it physically starts out in a linear combination of two different stationary states. The examples conclude the sum of the two wave functions has the same frequency as a photon. It is true that the numerical value of the correct frequency of the photon can be attained mathematically in this manner. However, it does not make physical sense. The so-called stationary states of motion are characterized by wave functions, but wave functions use the abstract characteristic frequencies for stationary states of motion, *not* the real physical electron oscillating frequencies of their motions. Also, the actual physical process of how the photon is created, and what causes its emission are not addressed in either of these examples. Because of this, many physicists believe there is something lacking in quantum mechanics. To clarify this subject, the rest of this section will provide a reasonable physical description of how an electron's physical quantum jump to a lower stationary state can produce a *causal* photon-creation process.

Now, let's consider how physically two oscillatory frequencies from different orbits can occur at the same time to cause the beat phenomenon. In order to solve the hydrogen atom two-body problem in quantum mechanics, the proton and electron are *mathematically* transformed into a one-body problem with the electron modified to have a reduced mass. This analysis requires the proton to be stationary in the calculations. However, in reality the two-body problem is *physically* still a two-body problem. The proton is not stationary. The electron and proton are each orbiting their center of mass, and the total energy of the atom includes the motion of the proton, as well as the electron. Thus, the energy of both the proton and electron must contribute to the energy of any emitted photon.

Under steady state conditions, regardless of the particular orbits, both the electron and proton must have the same orbital frequency. Furthermore, if the electron is oscillating while orbiting the proton, either the entire proton or a part of the proton must be oscillating at the same frequency, and in the polar opposite direction.

Therefore, I shall hypothesize the following:

Hypothesis VII:

The electron and proton in the hydrogen atom oscillate radially at the frequency $n_{orbit}f_{orbit}$, while in stationary orbits about their center of mass.

Now, if some disturbance causes an electron to jump from an upper orbit to a lower orbit, the energy of the atom will decrease, and the oscillatory frequency of the electron will increase. But since the proton will *not instantaneously* feel any effect from the electron's jump, it will at least retain its same momentum and oscillatory frequency during some small time lag. It will then try to catch up and regain its sync again with the electron's new stationary orbital and oscillatory frequencies. During that transition process the electron and proton will oscillate at different frequencies. The proton will retain the old frequency while the electron oscillates at the new frequency. Consequently, the two different oscillations of the electron and proton will radiate different frequencies in their combined edust field to cause the beat effect to form a wave that will have the frequency of the released photon. That combined transient edust flow field which propagates away during the change of both the electron and proton from one atomic steady state to another atomic steady state *is* the photon. The photon is caused by the combined motions of the electron *and* proton, *not* by two different states of the electron. We can now allay Rutherford's concern that Bohr's theory has a problem with lack of causality. The proton remaining momentarily at the old frequency is the reason the electron does not need to know its final stationary orbit before it gets there for the atom to emit the correct photon frequency.

Putting this in the form of a hypothesis:

Hypothesis VIII:

When the electron in a hydrogen atom moves from an upper to lower orbit, the frequency of the released photon is the quantity

ν = *(**electron** oscillatory frequency, i.e. orbital frequency times n oscillations per orbit in its new lower orbit, minus the **proton** oscillatory frequency while still in its old upper orbit)/2.*

*Furthermore, the photon released is the disturbed edust flow field resulting from the combination of the two oscillations during the transition period between the two different stationary **atomic** states.*

This seems natural since the lost mechanical energy from the atom is the total lost from the energy of the electron plus the energy lost from the proton.

Some commonly accepted physical theories imply or assume photons emitted from atoms are strictly massless, while others state massless but really only mean they don't have any rest mass. In any event, we can now conclusively state that photons emitted from atoms do have mass. They are transient waves of edust particles created from the combined motions of an electron and nucleus oscillating at different frequencies.

From this we can see examples of two different types of wave-particle dualities: (1) a single point charge with a high density of edust contained within the sphere of its radius, constituting a particle, and a field containing waves, if it is oscillating, made of edust particles propagating away or toward it, and (2) transient wave packets of edust particles, caused by electrons and nuclei changing stationary orbital states, which propagate away from atoms, and can generally act like single particles because of their short temporal and spatial duration. The historically envisioned wave-particle duality of a stationary or constant velocity point charge with an accompanying wave of a specific frequency and energy does not exist.

Bohr's Correspondence Principle is Obsolete

Bohr called the requirement that quantum physics give the same results as classical physics in the limit of large quantum numbers, or in the high energy states of the hydrogen atom, the *correspondence principle*. This was done because the quantum steps of the lowest-energy states of the electron in the hydrogen atom are relatively large, and classical reasoning failed. Bohr's correspondence principle which required the quantum mechanical photon frequency to match classical reasoning only in the large orbital radius limit is *now* unnecessary. Now with this new form of the Bohr theoretical Rydberg equation all photon frequencies have a valid classical interpretation from a beat frequency phenomenon. The photon frequency, both in large jumps as well as small jumps, is predicted from classical reasoning.

Theory Why Electrons Oscillate While Orbiting

From the characteristics of the electron at rest, or traveling at a constant velocity, there is no apparent reason to expect that the electron would demonstrate any wave characteristics. However, something appears to be making the electron and, consequently, the nucleus to oscillate radially while they orbit their center of mass in atoms. Now the question is: What causes the electron and nucleus to oscillate while in orbit? If the electron has a radial displacement amplitude with specific oscillation frequencies about the allowed orbits of the hydrogen atom, this must be a forced response to alternating directions of positive and negative fields coming from the proton. So, let's take a look at the composition of nuclei, the nucleons.

In experiments where electrons were scattered off of protons, protons were found to be a composite of so-called quarks as discussed in the following:

> ... electrons were deflected by the protons and the pattern of deflection told experimenters about the internal structure of the proton. ... There is no way to hold quarks in place while firing many electrons to map out the locations of the quarks. [22, p. 169-171]

Note that the locations of the quarks are not fixed, but move under the influence of the electrons.

Not only is the proton composed of quarks, but also the neutron is composed of quarks, and as discussed in the quote below quarks are charged particles:

> Although the neutron has no electrical charge overall, it contains charge within. As we shall see, both the proton and neutron have small but measurable extent within which are swirling electric charges, positives and negatives, ... known as quarks. [36, p. 69-70]

These quotes indicate that the structures of nucleons are quite active. The variability of the structures of nuclei, including the single proton in the hydrogen atom, seems to be supported by the fact that experimentally the proton has been shown to *not* have a fixed classical radius. The measurement of the proton's radius has been attempted in many experiments, and the measurement values vary. In fact, the radius is described as an *rms charge radius*, not as a classical fixed-value radius. When interacting with the electron, the charged particles inside of nucleons with their continuous motion and changing spatial locations may cause an orbiting electron or even an electron which just passes by to oscillate. Since an orbiting electron would certainly interact with the proton, the components of the proton could possibly adjust themselves to produce a different distribution of positive and negative forces to create the different number of spatial oscillations in each nth orbit.

From this analysis, it appears plausible that the distance and momentum of an electron with respect to nucleons, and their interplay may cause the electron to oscillate at different frequencies. If true, this could then also explain the outcome of electron diffraction experiments which suggested the wave-like nature of electrons. From this, I shall make the following hypothesis:

Hypothesis IX:

The interaction between electrons and the continuous motion and changing spatial locations of the charged particles inside of nucleons cause electrons to oscillate at different frequencies in different orbits, and also oscillate if they just pass by nucleons as in diffraction experiments.

Of course, if charged particles inside nucleons cause an electron to oscillate, the electron will induce the same oscillation frequency back into the nucleons.

Consequence for the Copenhagen Interpretation

Does a particle and its location exist without being measured, or must a measurement be made to collapse its wave function, allowing it to have a location?

According to David J. Griffiths in his book, *Introduction to Quantum Mechanics*:

> The collapse of the wave function is undoubtedly the *most* peculiar feature of this whole bizarre story. It was introduced on purely theoretical grounds, to account for the fact that an immediately repeated measurement reproduces the same value. [15, p. 431]

From the analysis above, it isn't surprising that an immediately repeated measurement reproduces the same value, and I conclude that orbiting and oscillating electrons exist without being measured. Also, since the wave function is only a mathematical means to determine the probability of finding a particle, it is not necessary for the wave function to collapse for the electron to have a specific location. The *Copenhagen interpretation* regarding the collapse of the wave function causing the existence of electrons is nonsense.

Chapter 7

The Mass Defect and Inadequacy of Coulomb's Law

It is a well known fact that the sum of the individual component nucleon masses of a nucleus is greater than the mass of the nucleus itself. This difference is known as the *mass defect*, and is currently thought to be caused by the release of energy from the constituent particles when a composite particle is formed, as discussed below:

> ... because of the contribution of energy to mass, the mass of a composite particle is not equal to the sum of the masses of its constituent particles. [2, p. 104]

> The fact that component masses do not add is a reminder that at the deepest level, a composite entity is not a simple combination of parts---it is a new entity entirely. ... Quantized mass is just there, awaiting explanation. [2, p. 105]

The above quotes illustrate the commonly accepted belief that somehow energy contributes to the mass of an object, and that no one knows how or why this hypothesized conversion process can occur. In addition to the explanation of quantized mass in Chapter 4, in this chapter quantized mass shall be explained for the mass defect, and it will be concluded there is no contribution of energy to mass. Furthermore, as a result of this, it will also be concluded that Coulomb's Law does not adequately characterize the forces between point charges at close distances.

The High Density of Nuclei

It has been concluded from empirical data plots and theoretical calculations that the density of all nuclei is the same, and that it is extremely high:

> ... we conclude that *all* nuclei have the *same density* of nuclear matter. It
> is easily found that the density of nuclear material is 2×10^{17} kg/m^3... .
> [31, p. 327]

This density is much greater than the average density of point charges at rest, which
is about 9.7×10^{12} kg/m^3, roughly by a factor of 20,000.

Now, because of the variable density nature of point charges, the density theoretically goes to infinity at their centers allowing for the possibility that a spherical
region near the center of point charges may account for this high nuclear density.
As shown in Figure 7-1, a small spherical volume at the center of an electron (and
positron) that has the same average density as nuclei has a radius of about 0.00697
times its radius at rest.

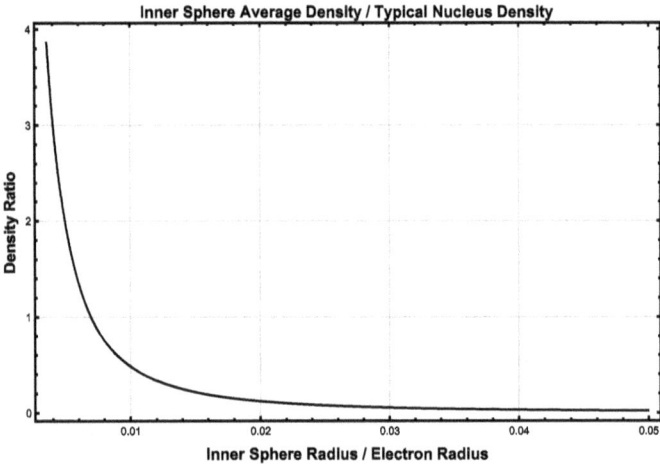

Figure 7-1. Ratio of Electron Inner Sphere Average Density to Typical
Nucleus Density vs. Ratio of Inner Sphere Radius to Electron Radius

If electrons and positrons could exist in nucleons where the centers of the point
charges were very close together, the average density of the nucleus of 2×10^{17}

kg/m^3 could be achieved. However, point charges may lose mass when they are close together, as nucleons do when they are near each other to combine to make a nucleus. If electrons and positrons, or the particles they form when close together are the components of nucleons, then perhaps they are the cause of the mass defect in nuclei.

To determine this we must ask the question: What makes all of the edust particles of a point charge at rest, and in motion, appear to hold together to make one single massive composite particle which has a total mass that can be measured as a property?

A Concept of Restraint Density

In Chapter 4 we saw that a sphere with the radius of an electron containing edust particles flowing in the radial direction, and causing a varying mass density field that decreased with increasing distance from its center, enclosed precisely the electron's rest mass. It seems logical to hypothesize that the reason an electron, when in its rest state, behaves like a single particle is that the density at its radius may be characterized as a *rest restraint density*. That is, a density that is large enough to significantly constrain the direction of motion of edust particles, while at smaller density they may freely stream in any direction away from the electron. This is somewhat analogous to a situation where people in a dense crowd must be part of the crowd until they reach the edge of the congestion, and can freely escape. Thus, I will make the following hypothesis:

Hypothesis X:

An edust density greater than or equal to the density at a point charge's radius at rest, called rest restraint density, is the cause for point charges to have a rest mass.

This would explain why an electron has a localized "particle" property and finite mass, but also has the properties of a field at the same time. It is both a finite mass of edust particles acting somewhat like a single particle, and also a fluid field of edust particles that theoretically can extend to infinity.

Electron-Positron Annihilation and Pair Production

In the previous section, I hypothesized that the positron and electron each have a rest restraint density at their radius which make them each behave as one composite particle. By choosing this density, we can draw the contours of these particles from the density field equation, i.e. equation (4-42), for any given separation distance. Furthermore, by merging the electron into the positron an electron-positron pair annihilation phenomenon can be simulated.

Assuming the annihilation occurs at a speed small compared to that of light, Figure 7-2 shows graphs of the rest restraint density cross-sectional contours of a positron and electron in Cartesian coordinates at seven different separation distances between their centers, a, beginning at a/r = 3, and ending at a/r = 0.001, where r is the radius of an electron. To facilitate a visual comparison, an eighth graph contains all of the individual previous seven graphs combined together.

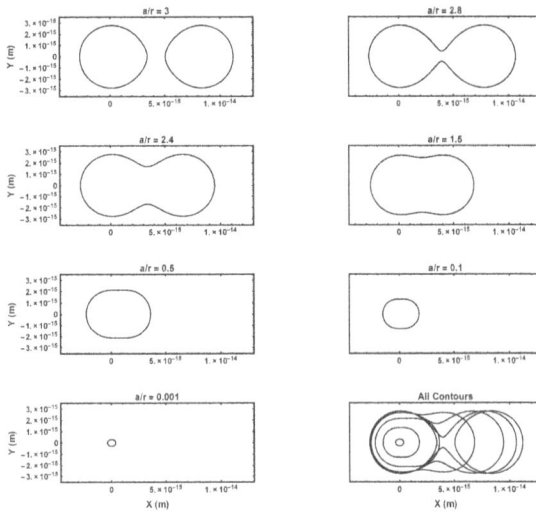

Figure 7-2. Rest Restraint Density Contours during Non-Relativistic Annihilation of a Positron and Electron.

It can clearly be seen that the two particles merge together, becoming one new particle, at slightly less than about three electron radii of separation. The new particle continues to change shape until it completely disappears if their centers coincide. This occurs because the mass flowing from the source has a progressively shorter path to traverse before it is absorbed by the sink. During this merging process the excess mass, once contained within the control volumes, generally propagates away as radiation in the form of two gamma rays. If the centers of the electron and positron could remain close without complete annihilation, perhaps by orbiting their center of mass, they may be separated in the reverse phenomenon called *pair production*.

Now, since the rest restraint density contours change shape as the positron and electron merge, let's determine the mass, volume, and average density of the new particles during this process.

Mass, Volume, and Average Density of New Particles

Integrations were performed over the rest restraint density contours as shown in Figure 7-2 at the various separation distances, as well as a few more, to obtain the masses, volumes, and average densities of the electron-positron pairs and merged particles. Values of the ratios of those quantities to the accepted measured rest values of a well separated electron-positron pair are listed in Table 7-1. These ratios are also plotted in Figures 7-3, 7-4, and 7-5, respectively.

As shown in Table 7-1, even when the positron and electron are separated by as much as five electron radii their total mass and volume are higher than their normally accepted measured rest values. This means that the "rest mass" and volume of electrons and positrons are *not* constants, but depend on their separation distance from other point charges. This also implies that the mass and volume of an *isolated* electron or positron cannot be measured exactly because they cannot be isolated. Just having a positron and electron in the same universe, even though they may be a very large distance apart, can cause each of their volumes and masses to increase.

Separation (a/r)	Mass Ratio	Volume Ratio	Density Ratio
5	1.00096	1.00208	0.998883
4	1.00257	1.00564	0.996951
3.5	1.00486	1.01077	0.994152
3	1.01127	1.02567	0.985956
2.8	1.01928	1.04576	0.974686
2.6	1.02939	1.06848	0.963416
2.4	1.03326	1.07065	0.965075
2.2	1.03033	1.05371	0.977812
2	1.01991	1.01331	1.00651
1.75	0.995062	0.952968	1.04417
1.7	0.988361	0.936932	1.05489
1.6	0.97308	0.902212	1.07855
1.5	0.955149	0.864107	1.10536
1.4	0.934378	0.822786	1.13563
1.2	0.883403	0.731172	1.2082
1	0.817935	0.628714	1.30097
0.5	0.565964	0.333764	1.6957
0.3	0.409578	0.203694	2.01075
0.1	0.186933	0.0687577	2.71872
0.001	0.00399709	0.000690081	5.79221

Table 7-1. The Ratios of Total Mass, Volume, and Average Density of an Electron-Positron Pair at given Separation Distances to their Total Accepted Measured Values.

It can also be seen in Figures 7-3 and 7-4 that as the positron and electron move closer together they increase total mass and volume until they reach a maximum somewhere between 2.2 and 2.6 electron radii of separation. When the separation distance is less than roughly a/r = 2.4 their total mass and volume continually decrease as they become closer.

It has been claimed in some books that a photon can convert into an electron-positron pair, or more dramatically stated that light can change into matter. The theories in this monograph dispute this. Photons are disturbances in the fields of streaming edust, and electrons and positrons are a type of source and sink. One cannot change into the other. Of course, this conclusion is probably made because the process cannot actually be seen, but only sensed by detectors. Perhaps, all that is known is that a photon goes into a "black box," and what is detected coming out or appearing is an electron-positron pair. However, we have seen that an electron-positron pair can be extremely small. The mass and volume decrease, and the opposite charges cancel, so they effectively become a small neutral undetectable particle. Suppose the electron-positron pair (i.e. positronium atom) can become a stable system at a very small size and not annihilate. Then it is easily understood that a high energy photon (which is only an electromagnetic field) could cause the positronium atom to be torn apart throwing the electron and positron in opposite directions, allowing each of their more massive fields to re-form, and allow them to be detected. There are no conversions of a photon into an electron-positron pair.

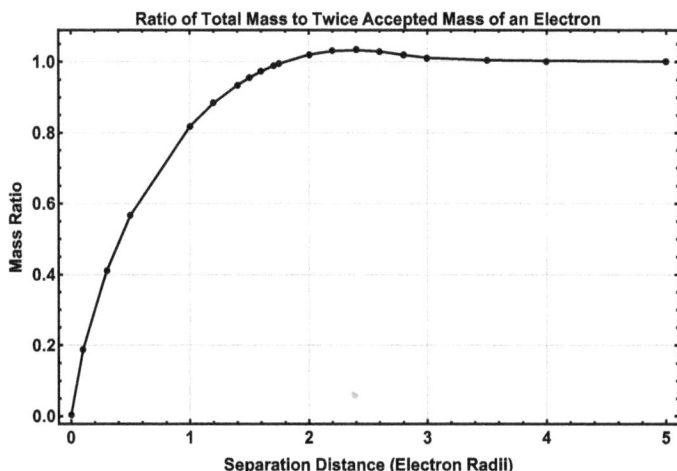

Figure 7-3. The Ratio of Total Mass of an Electron-Positron Pair at given Separation Distances to their Total Accepted Measured Values.

As shown in Figure 7-5, the average density trend does just the opposite as the mass and volume ratio curves. At a separation distance of five electron radii the average density is slightly below the "normal rest" average density of a point charge, continues to decrease slightly as the separation decreases until it reaches a minimum, and then reverses direction and increases apparently without bound. With the volume decreasing and density increasing, the total size and density of an accumulation of many pairs of oppositely charged point charges may approach the size and average density of the current theoretically predicted and measured value of nucleons, allowing the possibility that the sole components of nucleons are electrons and positrons.

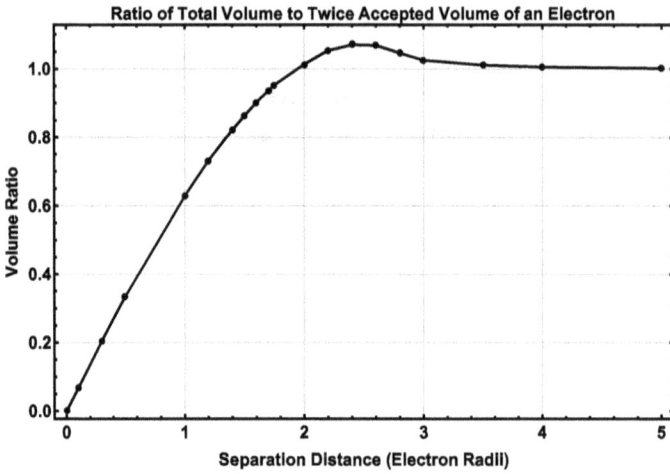

Figure 7-4. The Ratio of Total Volume of an Electron-Positron Pair at given Separation Distances to their Total Accepted Measured Values.

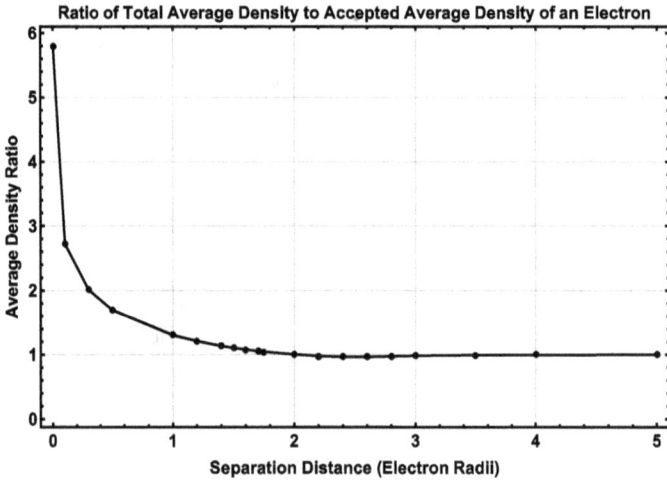

Figure 7-5. The Ratio of Average Density of an Electron-Positron Pair at given Separation Distances to their Average Accepted Measured Value.

As I have demonstrated, as an electron and positron closely approach each other their total mass decreases in the process. This same phenomenon occurs when nucleons form nuclei. If electrons and positrons are components of nucleons, their closeness may be the cause of the mass defect. Thus, I will make the following hypothesis:

Hypothesis XI:

Electrons and positrons are components of nucleons, and it is their closeness as nucleons move together that causes the phenomenon commonly referred to as mass defect.

In this process, there is no conversion of mass into energy, just the reduction of total edust mass, with their inherent kinetic energy, as the amount of the constrained edust field becomes smaller.

The Coalescing of a Positron Pair or an Electron Pair

We have seen that when an electron completely coalesces with a positron all of their mass is lost. If two like point charges are forced together, we would not expect that to occur. If we could push two positrons together so their centers coincide, we might expect that they would act like one positron with twice the mass flow rate. Under that circumstance we would expect the new double positron to have a larger radius than the single positron because its rest restraint density radius should increase. We can calculate the radius of a double positron if we set the density function of a positron from equation (4-6) with double the flow rate equal to the rest restraint density. That is,

$$\frac{2\,K}{4\,\pi\,c\,r_d{}^2} = \frac{K}{4\,\pi\,c\,r^2} \tag{7-1}$$

where r is the radius of a single positron, and r_d is the radius of the double positron. Solving for r_d, we find that

$$r_d = \sqrt{2}\ r \qquad\qquad\qquad (7\text{-}2)$$

Also, since we know the equation for the mass, (4-59), we can find the predicted mass, m_d, of the double positron in terms of a single positron, m_p.

$$m_d = \frac{2K\left(\sqrt{2}\ r\right)}{c} = 2\sqrt{2}\ m_p \qquad\qquad\qquad (7\text{-}3)$$

Now, given a predicted double positron radius and mass, we would like to see the details of what happens as the separation of like point charges decreases as they are forced to merge together.

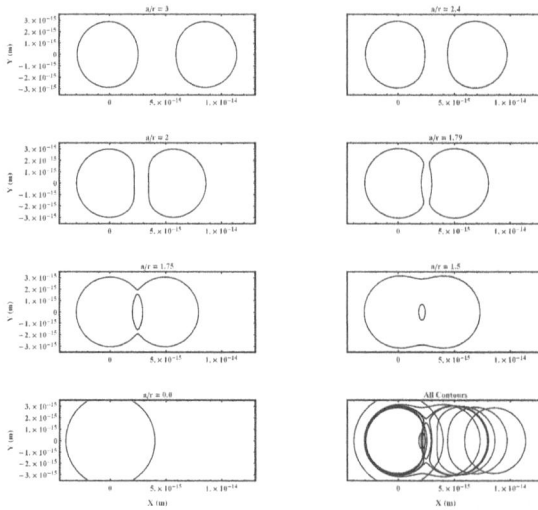

Figure 7-6. Rest Restraint Density Contours during Non-Relativistic Coalescing of a Positron-Positron Pair.

The rest restraint density contours of a positron-positron pair were computed at various values of separation beginning at $a/r = 3$, and ending at $a/r = 0$, and are shown in Figure 7-6. These contours are identical to those that would be attained from an electron-electron pair. We can roughly see that the numerical calculations confirm the predicted double positron radius in the plot where a/r = 0.

It is interesting that there are pockets of low density edust inside the outer rest restraint density contours on the plots for a/r = 1.5 and 1.75. The separation of a/r = 1.75 is approximately the point where the two point charges combine to form one particle. When this happens a field of edust density lower than the rest restraint density is contained inside, which continually becomes smaller as the two positrons continue to merge. A more detailed contour plot with various shades indicating different density values is shown in Figure 7-7 with the rest restraint density contours indicated by two closed thick black lines.

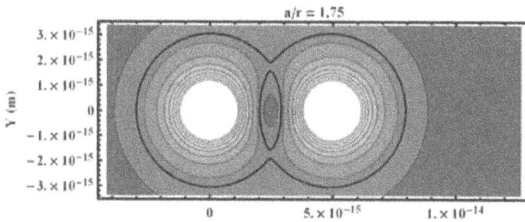

Figure 7-7. Mass Density Contours of a Positron-Positron Pair at a/r = 1.75.

Now, let's determine the mass, volume, and average density of the new particles during this like point charge merging process.

Mass, Volume, and Average Density of New Particles

Numerical integrations were performed over the rest restraint density contours shown in Figure 7-6, as well as a few more, to obtain the particles' masses, volumes, and average densities. Where the two positrons merged together there generally are outer and inner rest restraint density contours. Integrations of the mass, volume, and density over the volume enclosed by the outer contours are designated as the "Total" values. Since the inner volume is not strictly part of the new merged particle according to Restraint Density Theory, the inner masses and volumes were subtracted from the Total values to determine the "RDT" values of the new particles. The difference between the Total and RDT quantities is very small. Ratios of the Total and RDT values of mass, volume, and average density to the total accepted measured rest values of well separated positrons are listed in Table 7-2. These mass, volume, and average density ratios are also plotted in Figures 7-8, 7-9, and 7-10, respectively.

Separation (a/r)	Total Mass Ratio	Total Volume Ratio	Total Avg. Density Ratio	RDT Mass Ratio	RDT Volume Ratio	RDT Avg. Density Ratio
5	1.00091	1.00195	0.998954	1.00091	1.00195	0.998954
4	1.00229	1.00495	0.997357	1.00229	1.00495	0.997357
3	1.00767	1.01662	0.991197	1.00767	1.01662	0.991197
2.8	1.01027	1.02227	0.988259	1.01027	1.02227	0.988259
2.4	1.01982	1.04306	0.977718	1.01982	1.04306	0.977718
2.2	1.02887	1.06284	0.968039	1.02887	1.06284	0.968039
2	1.0441	1.09641	0.952289	1.0441	1.09641	0.952289
1.88	1.05903	1.12987	0.937298	1.05903	1.12987	0.937298
1.81	1.07184	1.15936	0.924511	1.07184	1.15936	0.924511
1.78	1.07921	1.17685	0.917038	1.07921	1.17685	0.917038
1.75	1.09654	1.22718	0.893549	1.09021	1.20449	0.90512
1.7	1.10901	1.25261	0.885365	1.10534	1.23901	0.892119
1.6	1.13038	1.28908	0.876893	1.12873	1.28278	0.879911
1.53	1.1444	1.30937	0.874009	1.1434	1.30553	0.875813
1.5	1.15027	1.31712	0.873322	1.14947	1.31401	0.874776
1.4	1.16942	1.33961	0.872955	1.16902	1.33807	0.873663
1.2	1.20642	1.37241	0.879051	1.20633	1.37206	0.879209
1	1.2423	1.39312	0.891742	1.24228	1.39305	0.891771
0.5	1.32913	1.41274	0.94082	1.32913	1.41274	0.94082
0.0	1.41421	1.41421	1.	1.41421	1.41421	1.

Table 7-2.   The Ratios of Total and RDT Mass, Volume, and Average Density of Two Positrons at given Separation Distances to their Total Accepted Measured Values.

We can see that the numerical calculations confirm the predicted double positron mass in Figure 7-8 where a/r equals zero. That is, from equation (7-3),

$$\frac{m_d}{2\,m_p} = \sqrt{2} \qquad\qquad (7\text{-}4)$$

This is also true for the volume ratio in Figure 7-9. Also, while both the mass and volume ratios vary from one at large separation to $\sqrt{2}$ at a/r = 0, the average density ratio in Figure 7-10 dips from one at large separation to a minimum of about 0.873 at a/r = 1.4, before returning back to one at a/r = 0.

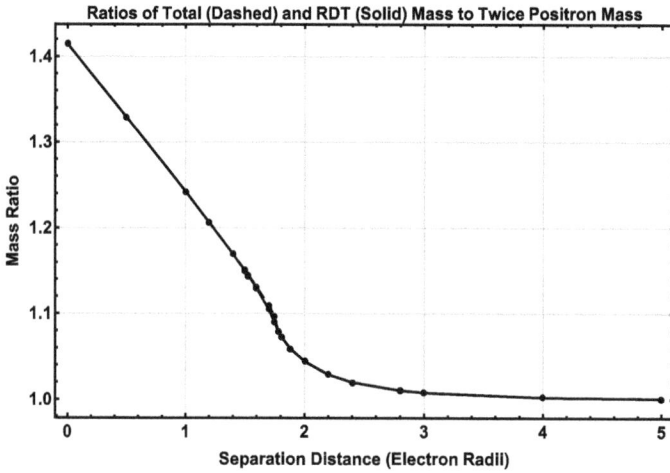

Figure 7-8. The Ratios of Total and RDT Mass of Two Positrons at given Separation Distances to their Total Accepted Measured Values.

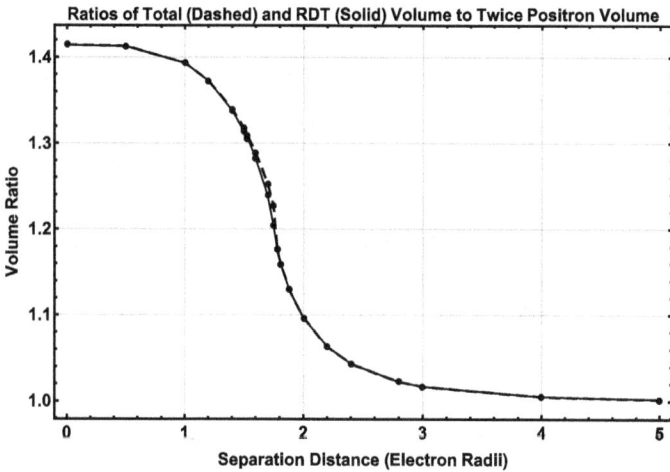

Figure 7-9. The Ratios of Total and RDT Volume of Two Positrons at given Separation Distances to their Total Accepted Measured Values.

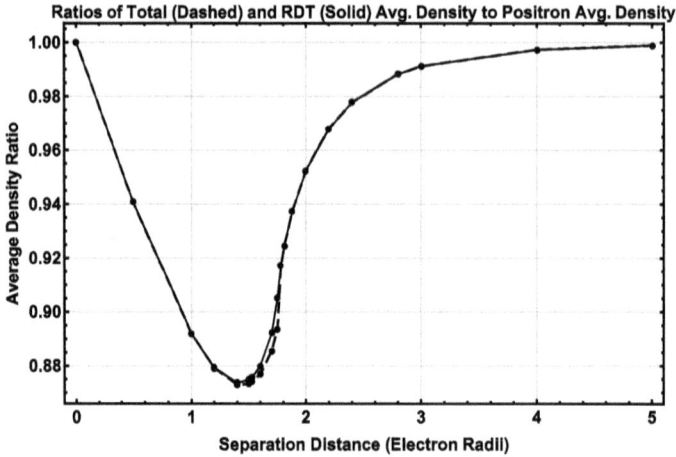

Figure 7-10. The Ratios of Total and RDT Average Densities of Two Positrons at given Separation Distances to their Average Accepted Measured Value.

Note that in Figures 7-3 and 7-8 both the electron-positron and positron-positron pairs increase mass as they approach each other from a large separation distance, but as they become very near the total mass of the electron-positron pair decreases and goes to zero while the mass of the positron-positron pair continues to increase to the maximum value of $\sqrt{2}$ times their total mass when widely separated. Also comparing Figures 7-5 and 7-10, the average density ratio of the electron-positron pair varies between slightly less than one to apparently increasing without bound near zero separation while the average density ratio of the positron-positron pair varies much less, and does not exceed the value of one.

Naturally, the electron-positron pair attract each other while the positron-positron pair repulse. So, a decrease in mass of the electron-positron pair is expected to occur naturally due to their own attraction, while the increase in mass requires an external force to cause the positron-positron pair to coalesce. If positrons and electrons are the sole constituents of nucleons there must be a balance between their separation distances, and attractive and repulsive forces to provide the experimentally measured values of mass defect and high density.

Divergence from Coulomb's Law

In Chapter 4, we calculated the force on a control volume containing a source which was caused by a sink. We found that the source-sink pair behaved like a positron-electron pair, and the force between them matched Coulomb's Law down to a separation distance of about 3 electron radii. Below that distance the force on the control volume diverged from the Coulomb force and decreased to zero as the two particles merged. However, we saw above that the shapes and densities of the fluid fields of the positron and electron begin to change significantly at a separation distance of roughly 3 electron radii. Assuming that Restraint Density Theory applies in this range, the objective of this section is to discover what forces occur between the newly shaped particles when they are closer than 3 electron radii.

Since we can calculate the density fields, vector fields, and rest restraint density contours, we should be able to calculate the forces on the new particles. Examples of the velocity vector fields and rest restraint density contours of a positron-electron pair and a positron-position pair, both with separation distances of a/r = 1.75, are shown in Figure 7-11.

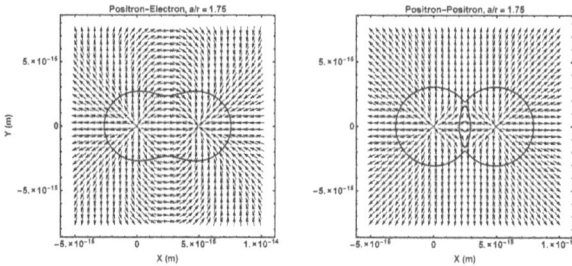

Figure 7-11. Velocity Vector Fields and Rest Restraint Density Contours of a Positron-Electron Pair and a Positron-Position Pair with Separation Distances of a/r = 1.75.

Forces were calculated on the new particles formed from positron-electron and positron-positron pairs with separation distances ranging from a/r = 0 to 5. Of course for the positron-electron pair, when a/r = 0 there is no longer any particle or particle pair for which an attractive force could be determined. In all of the other cases when any two particles merged into one particle the control surface was taken over one half of the particle. The control surface was constructed around the half-particle at the origin which included the common internal surface determined by a plane orthogonal to the x-axis at x = a/2, which cut the particle in half. In the case of two merged positrons, the force acting on this internal control surface between the two half-particles was zero due to symmetry.

The force of attraction between the positron and electron is shown with the corresponding Coulomb force in Figure 7-12. The force curve increases as the separation distance decreases similar to the Coulomb force until roughly a/r = 2.4, where it continues to increase, but at a slower rate. It also apparently continues to rise until a/r = 0 where the particles annihilate each other. The smallest separation distance plotted is at a/r = 0.001.

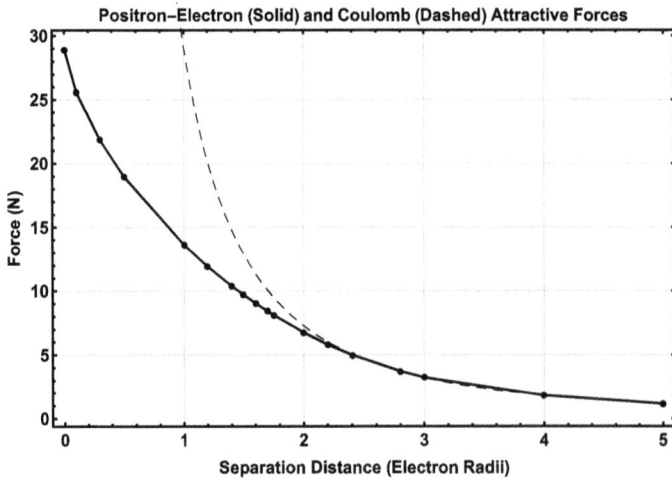

Positron–Electron (Solid) and Coulomb (Dashed) Attractive Forces

Figure 7-12. The Force of Attraction between a Positron and Electron (Solid) with the Corresponding Coulomb Force (Dashed).

The force of repulsion between two positrons (or two electrons) is shown with the corresponding Coulomb force in Figure 7-13. This force curve diverges from the Coulomb force at roughly a/r = 1.75, and levels off to a repulsive force of about

-14.5268 newtons as the separation approaches zero. Of course, if the two positrons could completely merge to become one double positron they would then have no force at all between them.

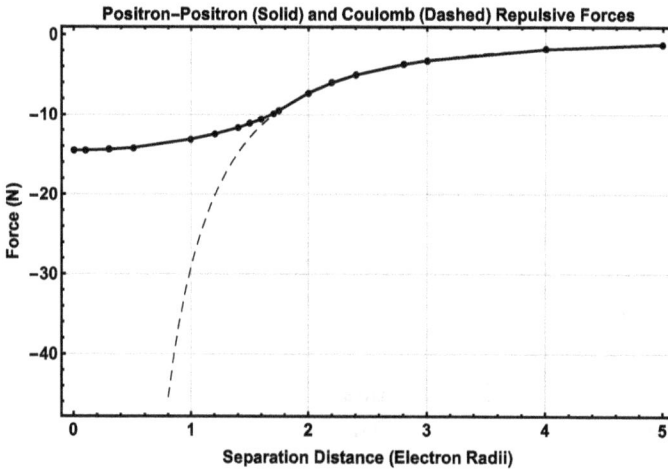

Positron–Positron (Solid) and Coulomb (Dashed) Repulsive Forces

Figure 7-13. The Force of Repulsion between Two Positrons (or Two Electrons) (Solid) with the Corresponding Coulomb Force (Dashed).

It is interesting to see a comparison of the characteristics of the attractive and repulsive forces plotted together in Figure 7-14. They behave in different manners, and neither is in accordance with Coulomb's Law.

The attractive and repulsive force curves are added together in Figure 7-15 to further show the difference between the electrostatic attractive and repulsive forces as the two particles merge. The smallest separation distance plotted is at $a/r = 0.001$.

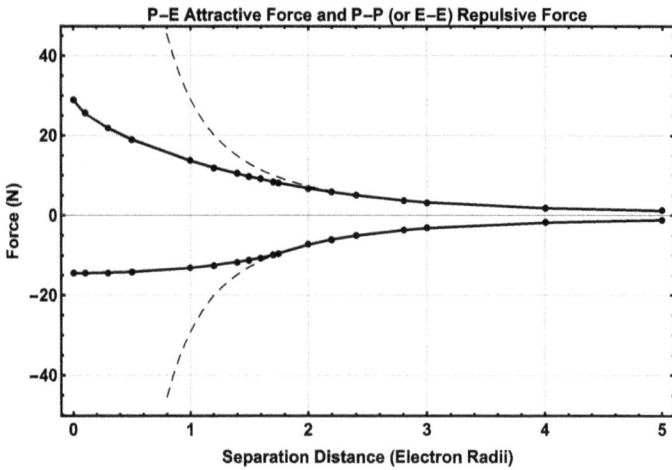

Figure 7-14.   Electrostatic Attractive and Repulsive Forces when Two
Point Charges are Very Close Together, Coulomb Forces (Dashed).

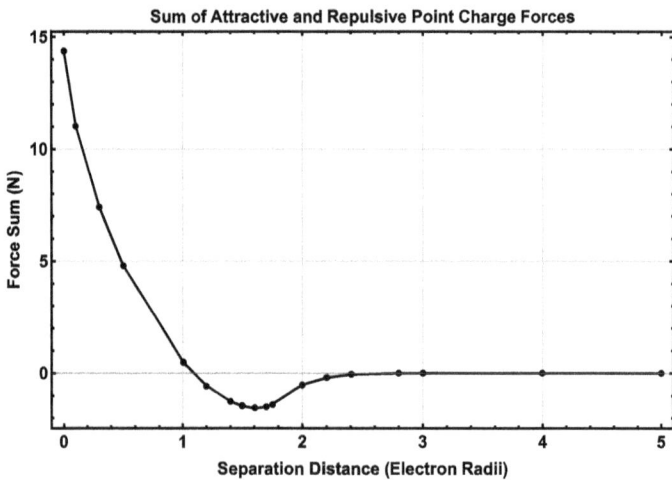

Figure 7-15.   Sum of Attractive and Repulsive Electrostatic Forces.

From this analysis it is concluded that the electrostatic forces on point charges
diverge significantly from Coulomb's empirical force law as they merge at small

separation distances. Also, the absolute value of the magnitude of the attractive and repulsive forces are generally not the same strength in this region. At a separation distance less than roughly 1.1 electron radii the attractive force exceeds the repulsive force, and at greater than roughly 1.1 electron radii the absolute value of the magnitude of the repulsive force exceeds the attractive force.

A listing of the numerical values of the attractive, repulsive, and Coulomb forces computed for the plots can be found in Table 7-3.

Separation (a/r)	P-E Force (N)	P-P Force (N)	|Coulomb Force|(N)
5	1.16214	-1.16214	1.16214
4	1.81584	-1.81584	1.81584
3	3.22817	-3.22817	3.22817
2.8	3.70579	-3.7058	3.7058
2.4	4.98802	-5.04401	5.04401
2.2	5.80372	-6.00279	6.00279
2	6.74294	-7.26338	7.26338
1.75	8.10136	-9.48319	9.48686
1.7	8.39897	-9.90469	10.0531
1.6	9.02143	-10.585	11.349
1.5	9.68147	-11.1529	12.9127
1.4	10.3808	-11.6462	14.8232
1.2	11.9061	-12.47	20.176
1	13.6188	-13.1249	29.0535
0.5	18.9853	-14.1843	116.214
0.3	21.8114	-14.4039	322.817
0.1	25.5417	-14.5131	2905.35
0.001	28.8901	-14.5268	2.90535×10^7
0	undefined	0.	∞

Table 7-3. Listing of Attractive, Repulsive, and Corresponding Coulomb Electrostatic Forces at Point Charge Separation Distances.

Does "Pure Energy" Exist?

I have seen and heard the phrase "pure energy" frequently used in the media, but not defined. Energy has always been defined with respect to mass, either as kinetic energy or potential energy. However, radiation is said to have energy, but be massless, and has been referred to as "pure energy." Still, in the previous chapter it was concluded that radiation is not massless. It is composed of mass with kinetic energy. Also, it has been said that "kinetic energy" is able to convert into "mass energy" [2, p. 19-21], whatever that is. Furthermore, from the first quote in this chapter, it is said that "... because of the contribution of energy to mass, the mass of a composite particle is not equal to the sum of the masses of its constituent particles" [2, p. 104]. However, energy does not contribute anything to mass. Mass *with* kinetic energy is simply lost or gained when two oppositely charged composite particles become closer or separate from each other. Another interpretation of the relationship between mass and energy comes from Einstein's equation, $E = m_0c^2$. The interpretation often given to this equation is that mass and energy are equivalent and interchangeable, i.e. one form can convert into the other. But again, mass is *not* energy. A simple look at the units of each will show they are not the same. Furthermore, the inherent internal energy of a mass at rest, such as a point charge, has been said to be "pure energy," but rest mass only contains edust particles moving at the speed of light. That internal energy is the kinetic energy of the edust particles. Therefore, "pure energy" and the supposed phenomenon that energy can convert into mass do not appear to exist, and scientists and authors should refrain from referring to them.

Chapter 8

Is Matter Composed Solely of Electrons and Positrons?

If the edust particle is *the* fundamental particle of the universe, and it is created and destroyed, or otherwise enters and exits, by way of positrons and electrons, are those flow patterns the only means for this to occur? If they are, then all matter is fundamentally composed of the composite particles, positrons and electrons, and this provides the explanation for the existence of all known macroscopic particle masses. However, can electrons and positrons exist inside nuclei and nucleons?

Can Point Charges Exist Inside Nuclei and Nucleons?

In the history of physics, electrons have both been thought to exist and not exist inside nuclei. Now, it is a well known experimental fact that electrons are emitted by unstable nuclei in a process called beta decay:

> ... beta decay, in which an electron is observed to be emitted from the nucleus ... [7, p. 68]

> ... an unstable nucleus may emit photons or particles of non-zero rest mass (for example, ... β-particles). [31, p. 334]

After Ernest Rutherford discovered that beta radiation (later to be shown to consist of electrons) was emitted as a result of the radioactivity of uranium, it was proposed that the emitted electrons did not pre-exist in the atom, but were created during the process from the energy released [37, p.10]. However, there is no reason now to believe that electrons are created from "energy." It is likely they exist inside nuclei.

It is also said that when a star collapses, protons and electrons combine to form neutrons [37, p.139]. Thus, from this it is clear that a neutron contains at least one electron. So, it appears electrons must exist in some nuclei.

I suppose the current resistance for nuclei to contain electrons is due to concepts in quantum theory as discussed below:

> Electrons shooting from the nucleus were unwelcome ... . Quantum mechanical theory says that an electron *can't* be confined within a nucleus. ... would have a fairly defined position ... uncertainty of momentum would be large ... so much kinetic energy that it would fly out ... [2, p. 39]

Also, it has been stated that electrons *cannot* exist inside nuclei even while stating they are emitted from nuclei. This was justified by the belief that the electron must be excluded because of the size of its accompanying de Broglie wavelength:

> We are now in a position to discuss a question that has not been raised to this point; namely, why electrons do not exist as constituents of nuclei. Before the discovery of the neutron in 1932, it was thought that nuclei consisted of protons and electrons. ... The hypothesis that electrons existed in nuclei was strengthened by the observation that radioactive materials undergoing β^-- decay actually emit electrons from their nuclei.
> If an electron were confined and localized to within a nuclear dimension, say, 10^{-15}m, then the de Broglie wavelength of the electron could be no greater than this distance. ... it must, therefore, be concluded that electrons cannot exist within a nucleus. [31, p. 327]

However, from the theory above in Chapter 6, it was concluded the de Broglie wave does not exist as an accompanying wave to an electron as it was historically hypothesized. So, that objection no longer applies.

It is also said that electron-positron pair production can be caused by photons disturbing high atomic number nuclei. This indicates that small radii positronium atoms may be part of nucleons. In fact, electrons and positrons are said to be found in the debris from proton-antiproton collisions in particle accelerators:

> ... when a proton meets an antiproton. ... The faster they collide, the greater their energies, the more pions or gamma rays are produced ... the pions, short-lived constructs of a quark and an antiquark, self-destruct, turning into yet more gamma rays, or into electrons, positrons, and ghostly neutrinos, ... [36, p. 78-79]

In addition to the self-destruction of quark-antiquark pairs into electrons and positrons, apparently the reverse can occur:

> One of the definitive experiments which supports the quark model is the high energy annihilation of electrons and positrons. The annihilation can produce muon-antimuon pairs or quark-antiquark pairs which in turn produce hadrons. The hadron events are evidence of quark production. [38]

This essentially says that the collisions of solely positrons and electrons can create the components of nuclei.

Addressing the Fractional Charge of Quarks

The subatomic so-called "elementary" particles called quarks are said to have the property of fractional electric charges, and be bound together by a specific "elementary" particle ("force carrier"), called gluons, as briefly discussed below:

> Quarks are point-like, spin-1/2 particles, with electric charges of $\pm$ 1/3 or $\pm$ 2/3, that are bound together by the gluons of QCD. Still, neither quarks nor gluons have ever been directly observed as tracks in a cloud chamber or particle detector. [22, p. 182]

Even with this unusual situation, it seems reasonable to expect that sources and sinks are the only fundamental flow patterns of edust causing the creation of matter. The possibility exists that some sources and sinks may have mass flow rates different from positrons and electrons, but fractional charges have never been directly observed. If there were different flow rates, we would expect there would be equal numbers of each positive and negative fractional flow rate sources and sinks, as well as whole numbered flow rates in the universe to preserve a global conservation of equal positive and negative charge, or a neutrally charged universe. Fractionally charged point charges are perhaps an unnecessary complication when there is believed to be a simpler answer.

The existence of "fractional charge" was explained in 1983 by R. B. Laughlin [39], and in 1998 Robert Laughlin, Horst Störmer, and Daniel Tsui were awarded the Nobel Prize in Physics for the discovery and explanation of the fractional Hall effect which attributes electrons as the cause of forming quasiparticles. Below is a convincing excerpt from Horst Störmer's Nobel lecture:

> The fractional quantum Hall effect is a very counterintuitive physical phenomenon. It implies that many electrons, acting in concert, can create new particles having a charge *smaller* than the charge of any individual electron. ... But fractional charges are very bizarre indeed. Not only are they smaller than the charge of any constituent electron, but they are exactly 1/3 or 1/5 or 1/7 etc. of an electronic charge, depending on the conditions under which they have been prepared. And yet we know with certainty that none of these electrons has split up into pieces. [40]

Furthermore, Lee Smolin has correctly stated that:

> ... quarks, electrons, and neutrinos are just different manifestations of the same underlying kind of particle. ... there have to be new physical processes by which they can turn into one another. [5, p. 63]

The theory herein has already explained how point charges can create new particles which have a smaller field structure, and cause a smaller force on other objects.

Thus, from this discussion it seems reasonable to conclude that electrons and positrons may be the constituents of nucleons, and appear to be the only fundamental flow patterns causing the existence of matter.

Equal Matter and Antimatter?

Since nucleons appear to consist of electrons and positrons in nearly equal parts, and with the orbiting electrons the atoms are electrically neutral, all antimatter said to be formed at the moment of the so-called Big Bang may still exist. It may be that when an electron and positron collide they do not completely destroy each other even though most or all of the edust within their control volumes is emitted as radiation. The entire mass of edust within the restraint density surface may be released/radiated, but the flow may still continue between the point charge centers which may allow them to re-form again in pair production. The merged particle will appear to be uncharged, nearly invisible, and most likely impossible to detect.

This would also answer the question of how a matter-dominated world emerged from the symmetric matter-antimatter universe that is said to have been produced in the Big Bang [37, p. 147]. It is said that after creation and/or the Big Bang:

> ... energy coagulated into matter and its mysterious opposite, antimatter, in perfect counterbalance. ... Today antimatter does not exist normally, at least on earth, a vanishing act that is one of the unexplained mysteries of the universe. [36, p. x]

Here it is believed that there used to be equal quantities of matter and anti-matter, but perhaps no longer. As already discussed, I do not believe that energy "coagulated" into matter. Also, I do believe antimatter may exist normally on earth. What we call neutrally charged matter is really half antimatter, perhaps in extremely close orbital configurations. Below is an interesting account by author Frank Close of the first demonstration when matter and antimatter were said to be caused by radiation:

> Many of their pictures showed up to twenty particle tracks diverging from

some point in a copper plate just above the chamber like water from a shower. ... showing that roughly half of the particles were negatively charged and the rest positively charged. Blackett and Occhialini realized that as positrons do not occur naturally on earth, the appearance of equal numbers of positrons and electrons must be because they were being produced by some invisible high energy cosmic radiation. The message was that the positrons were being formed as a result of collisions between the cosmic rays and the atoms in the chamber. [36, p. 58]

However, now all neutrally charged matter appears to naturally consist exclusively of pairs of electrons and positrons. If this is true, then we can conclude that there is an equal amount of matter and antimatter in the universe as expected, but not previously understood, and antimatter is not mysteriously created on earth, but caused to be detectable in pair production.

What is Matter?

This chapter has provided much evidence that electrons and positrons exist as the sole components of particles commonly referred to as matter. And, although this chapter has not conclusively determined that all matter is composed solely of electrons and positrons, at this point there is no reason not to believe it. Therefore, this seems like a reasonable hypothesis for further examination.

Hypothesis XII:

All matter is fundamentally composed solely of the flow patterns of positrons and electrons, which provide the only means of creation or entry, and destruction or exit, of edust particles in our three-dimensional universe.

If this is true, it leads to plausible definitions of the terms matter and radiation, which have previously not been well defined.

Proposed Definitions of Matter and Radiation

Historically, the constituents of our universe have been thought to be matter and radiation, where matter consisted of mass and radiation did not. Since we now know that all matter and radiation are composed of mass, i.e. edust particles, the results above lead to a proposal of their definitions. Since electrons and positrons are hypothesized to be the sole contributors to the mass of macroscopic particles commonly called matter, in the simple definitions below matter is defined to be synonymous with any object that contains at least one of them.

Proposed Definition of Matter:

Matter is any object that contains at least one point charge.

Note here that antimatter is included in the definition of matter since it *causes* an accumulation of edust with high density, and contributes to the mass of macroscopic particles. Furthermore, we don't know whether the positron or electron is a source or sink to really distinguish which one produces edust and which one eliminates it. If we did, then perhaps we may want to more definitively distinguish between which is matter and which is antimatter than the current convention.

Also, matter is composed of composites of restrained particles. Individual edust particles, or pure radiation, have been excluded from the definition of matter even though they have mass. That is, all mass is not matter.

Proposed Definition of Radiation:

Radiation is any object that does not contain any point charges, i.e. radiation is strictly only edust, and does not contain anything that produces or eliminates edust.

Note that while radiation alone may perhaps accumulate to have a high density, perhaps higher than the rest restraint density, the process by which this may occur is not the same as those when created or destroyed by point charges. Also, in accordance with this definition there *is* constant radiation from point charges at rest. Radiation is edust flow whether it is oscillating with some frequency, or not.

Chapter 9

Does Edust Theory Comply with Relativistic Phenomena?

Let me begin by noting that none of the derivations so far have involved any relativity principles because the point charges examined were either fixed or assumed to be moving slowly in an inertial frame of reference. The only rapidly moving particles were edust particles moving at the maximum speed of light. However, now I will extend the theory to see if it complies with relativistic phenomena. We will examine point charges moving with a velocity near the speed of light with respect to a fixed inertial reference frame, determine how their mass density patterns change with speed, and calculate their relativistic mass and kinetic energy.

Relativistic Mass Density of a Point Charge

From the relativistic mechanics of continua it can be shown that the *relativistic mass density* of a fluid continuum in motion is different from the mass density, ρ, while at rest. For a fluid continuum subject to no forces other than mutual collisions, the equation for the relativistic mass density, ρ_{rel}, for every point in a flow where the infinitesimal fluid elements can be treated in the spirit of a particle moving with speed v, is

$$\rho_{rel} = \gamma \rho \qquad\qquad (9\text{-}1)$$

[41, p. 178-181], where

$$\gamma = \frac{1}{\sqrt{1-(v^2/c^2)}} \qquad\qquad (9\text{-}2)$$

This phenomenon applies to the density of mass in an electromagnetic field which can be considered as moving with the speed v. Also, we know from Chapter 5 that mass density at rest is proportional to electric field strength as given by equation (5-7), repeated below.

$$\rho = \frac{\sqrt{K/Ck}}{4\pi c} E \tag{5-7}$$

Thus, substituting equation (5-7) into (9-1) we have

$$\rho_{rel} = \gamma \frac{\sqrt{K/Ck}}{4\pi c} E \tag{9-3}$$

That is, we have the equation for relativistic mass density provided the infinitesimal fluid elements that make up the electric field E can be treated in the spirit of a particle moving with speed v.

Now, it is historically well known that the electric field of a point charge has a relativistic form where it can be considered to move with the speed v. The electric field strength, E_{rel}, of a uniformly moving point charge at speed v has been derived in several books on the special theory of relativity, e.g. [23, p. 461][42, p. 116]. The equation for the relativistic electric field strength in the radial direction of a moving point charge relative to its *present* location in an inertial reference frame is given by

$$E_{rel} = \frac{e}{4\pi\epsilon_0\, \gamma^2\, R^2\left[1-\left(v^2/c^2\right)\sin^2\theta\right]^{3/2}} \tag{9-4}$$

Substituting equation (9-4) into (9-3), we have

$$\rho_{rel} = \gamma \frac{\sqrt{K/Ck}}{4\pi c}\, \frac{e}{4\pi\epsilon_0\, \gamma^2\, R^2\left[1-\left(v^2/c^2\right)\sin^2\theta\right]^{3/2}} \tag{9-5}$$

Now, taking the value of e, as derived in equation (5-5),

$$e = \sqrt{\frac{CK}{k}} \tag{5-5}$$

and substituting it into equation (9-5), we have

$$\rho_{rel} = \gamma \, \frac{\sqrt{K/Ck}}{4\pi c} \, \frac{\sqrt{CK/k}}{4\pi \epsilon_0 \, \gamma^2 \, R^2 \left[1-\left(v^2/c^2\right)\sin^2\theta\right]^{3/2}} \tag{9-6}$$

After consolidating terms, the relativistic mass density function, ρ_{rel}, of a point charge moving with speed v relative to an inertial reference frame, in terms of the mass flow rate constant K, becomes

$$\rho_{rel} = \frac{K}{4\pi c\,\gamma R^2 \left[1-\left(v^2/c^2\right)\sin^2\theta\right]^{3/2}} \tag{9-7}$$

As a check, if we set v = 0 in this expression, we get the mass density function, equation (4-6), for a stationary point charge, as expected.

$$\rho = \frac{K}{4\pi c\,R^2} \tag{4-6}$$

Now that I have derived the relativistic mass density function for a point charge, it is of interest to plot it for several different ratios of velocity to the speed of light. In Figure 9-1 are density and density contour plots for v/c = 0.0, 0.7, 0.9, and 0.97 where the direction of velocity is to the right.

We can easily see that the mass density fields change shape with speed. The spherical shape at rest flattens in the direction of motion. This results in a roughly oval-like or pinched oval-like shape as viewed perpendicular to the motion. The field also expands outside the rest control volume in the direction perpendicular to the axis of motion while maintaining circular symmetry about that axis. Furthermore, at very high speed the shape becomes significantly thinner near the center than at other parts of the point charge in the direction of motion.

Of course, the edust particles do not move at speed v with the point charges. They always move at the speed of light whether being a constituent of a point charge at rest or in motion. What we measure as point charge movements are high density concentrations of edust changing positions in the field.

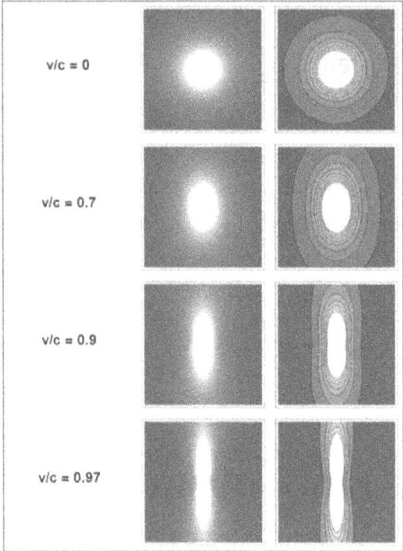

Figure 9-1. Relativistic mass density and density contour plots for a point charge moving with speed v relative to a fixed inertial reference frame where the direction of velocity is to the right.

Now that I have derived the relativistic mass density function for a point charge, and plotted it for various values of velocity, it is of interest to determine a point charge's *relativistic mass*.

Although Einstein derived equations for longitudinal and transverse mass in his early papers [43, p. 414], he later expressed dislike for the concept of relativistic mass which he felt could be given "no clear definition":

It is not good to introduce the concept of the mass $M = m/(1 - v^2/c^2)^{1/2}$ of a moving body for which no clear definition can be given. It is better to introduce no other mass concept than the "rest mass" m.

Albert Einstein in a letter to Lincoln Barnett, 19 June 1948 [44]

It is now time to propose a "clear definition" for relativistic mass.

Relativistic Restraint Density Theory

As we saw in Figure 9-1, as a point charge increases speed its density field contours flatten into a circular disk with an oval or pinched oval-like cross-section. If the concept of rest restraint density is valid for an electron at rest, it seems reasonable that the cause for relativistic mass may be due to a Relativistic Restraint Density.

Let us now derive a *relativistic restraint density function* from the *rest restraint density*. We know from equation (9-1) that the relationship between the relativistic density and rest density of a continuum is

$$\rho_{rel} = \gamma \rho \qquad\qquad (9\text{-}1)$$

For a point charge at rest, the density at its radius *is* its *rest restraint density*. Thus, by substituting equation (4-6) into (9-1) with R changed to the electron's rest radius, r, its relativistic restraint density becomes

$$\rho_{rel-restraint} = \gamma \, \frac{K}{4\pi c r^2} \qquad\qquad (9\text{-}8)$$

Now setting the relativistic restraint density equal to the relativistic density function of a point charge, equation (9-7), we have

$$\gamma \, \frac{K}{4\pi c r^2} = \frac{K}{4\pi c \gamma R^2 \left[1-\left(v^2/c^2\right)\sin^2\theta\right]^{3/2}} \qquad\qquad (9\text{-}9)$$

where now R = r_{rr} is the relativistic restraint radius function for a point charge.

Solving for this function, we have

$$r_{rr} = \frac{r}{\gamma\left[1-\left(v^2/c^2\right)\sin^2\theta\right]^{3/4}} \tag{9-10}$$

Polar plots of the rest restraint radius of a point charge for v/c = 0, and relativistic restraint radius function of a point charge for v/c = 0.97 are shown in Figure 9-2. Also shown is the Lorentz-Fitzgerald contraction of the electron's circular cross-section at v/c = 0.97.

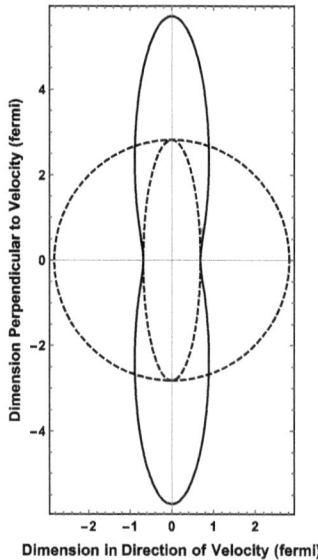

Figure 9-2. Polar plots of Electron Radius at v/c = 0 and its Lorentz-Fitzgerald Contraction at v/c = 0.97 (both Dashed), and the Relativistic Restraint Radii at v/c = 0.97 (Solid).

The pinched oval-like cross-sectional shape of the relativistic restraint radius function plot for v/c = 0.97 has the same shape as the previous corresponding density plot in Figure 9-1. Also, the relativistic restraint radius function has the same contraction factor in the direction of motion at $\theta = 0$ as the Lorentz-Fitzgerald contraction factor, $1/\gamma = \left(1 - v^2/c^2\right)^{1/2}$. Consequently, this derivation provides the *first* fluid dynamics justification for the original hypothesis by H. A. Lorentz that the form of the electron experiences a physical contraction in the direction of motion. However, in this field theory, the thickness of the point charge in the direction of motion is *larger* at some points than predicted by the contraction factor, and in the perpendicular direction the electron's height is *increased* by a factor of $\left(1 - v^2/c^2\right)^{-1/4}$. This is surprising because a law of increased height in the perpendicular dimensions was said to be impossible:

> There *cannot* be a law of contraction (or expansion) of perpendicular dimensions, for it would lead to irreconcilably inconsistent predictions. [23, p. 518]

This does *not* indicate any problem with the Lorentz-Fitzgerald contraction or the Lorentz coordinate transformation group. Both the *geometrical shape* of a body, which is the shape measured with a system of coordinate axes which are at rest with respect to it, and its *kinematic shape*, which is historically defined as the coordinates which express the simultaneous positions of its various at-rest boundary points when it is in motion with respect to the axes of reference, are different from its *relativistic restraint density* shape. The density of the field just changes because of the contraction which then results in a new restraint density boundary for the moving point charge. This is clearly illustrated by the three shapes in Figure 9-2.

Perhaps these two new predicted phenomena, variable width and increased height boundaries of matter, can be tested in the future. Several examples of thought experiments of high speed macroscopic objects attempting to fit into smaller spaces than their rest dimensions, typically presented in textbooks concerning relativity (e.g. a fast car becoming shorter to fit in a garage), apparently may need to take these phenomena into consideration so accidental scraps and collisions between the two objects do not occur.

It is now time to determine if Relativistic Restraint Density Theory predicts the correct value of relativistic mass, but first I need to relate a very abbreviated summary of a small part of the historical development of relativistic mass. A more extensive history of this subject is well documented in the literature.

Short Historical Development of Relativistic Mass

It was recognized as early as 1818 that J. J. Thomson believed in the concept of electromagnetic inertia. However, the original idea of mass varying with speed was not introduced until 1899 in a paper by H. A. Lorentz [45]. This was followed by research on this topic by a long list of scientists, including Walter Kaufmann (1901), Max Abraham (1902), Alfred Bucherer (1904), Paul Langevin (1905), Albert Einstein (1905), Max Planck (1906), Henri Poincaré (1906), Hermann Minkowski (1908), G. N. Lewis (1908), and R. C. Tolman (1909).

After the paper by Lorentz, in 1902 Abraham derived a formula for the electromagnetic mass of moving bodies from the concept that the mass of a charged particle is associated with its electromagnetic properties. After this derivation, he concluded that his equation for mass was only valid in the longitudinal direction, i.e., the force causing the motion is in the direction of motion, and that the electromagnetic mass also depends on the direction of the moving body with respect to the forces acting on it. Thus, Abraham also derived another equation for mass where the force is in the transverse direction to its motion. Abraham was apparently the first to use the expressions of "transverse" and "longitudinal" masses. Unfortunately, his theory suffered from some serious difficulties, mainly due to his model of the electron as a rigid sphere with a "charged" surface. The rigid body does not allow for the Lorentz-Fitzgerald contraction.

In the 1899 paper, Lorentz had assumed that electrons undergo length contraction in the line of motion, which led to results for the acceleration of moving electrons that were different from those given by Abraham. Lorentz expanded his 1899 ideas in his famous 1904 paper, where he developed his own equations for longitudinal and transverse mass. These equations can be put in terms of the rest mass, m_0, and γ as

$$\text{Longitudinal mass} = m_0\gamma^3 \qquad (9\text{-}11)$$

$$\text{Transverse mass} = m_0\gamma \qquad (9\text{-}12)$$

In 1901, Kaufmann had already conducted experiments to measure the mass of electrons at relativistic speeds. In those experiments, the force was perpendicular to the velocity, so the transverse equations for mass of Abraham and Lorentz could be tested by Kaufmann's data. Unfortunately, Kaufmann's data could not unequivocally determine which of the competing theories was best. Later in 1940, an experiment by Rogers et al. produced more accurate data that more closely agreed with Lorentz [46, p. 230]. Apparently, no experiments that could measure the longitudinal mass were conducted, even though the mass was predicted to increase at a much greater rate, i.e. vary as γ^3.

In 1905, Einstein derived similar equations of motion to that of Lorentz by assuming "a point particle possessing an electric charge" called an electron [43, p. 412]. From these equations of motion Einstein determined his own equations for longitudinal and transverse mass as

$$\text{Longitudinal mass} = m_o\gamma^3 \tag{9-13}$$

$$\text{Transverse mass} = m_o\gamma^2 \tag{9-14}$$

Einstein's longitudinal mass was identical to that of Lorentz, but his transverse mass was different by an extra factor of γ. Even though the two equations for transverse mass were different, and Lorentz's equation fit Kaufmann's data best, Lorentz's equation (9-12) became known as the Lorentz-Einstein equation.

In 1909, Lewis and Tolman [47][48, p. 37], published a thought experiment without using electromagnetic theory at all. It is commonly used in textbooks today. It involves a relativistic collision of two spheres using only conservation laws and the postulates of the special theory of relativity. The resulting equation was identical to the Lorentz-Einstein (L-E) equation for transverse mass, and Tolman claimed it is the "general expression" for the mass of a moving body. Because the two equations were identical, it was apparently immediately accepted as the valid equation for relativistic mass.

It is now of interest to compare the relativistic mass from the L-E equation to the relativistic mass enclosed within a freely moving point charge's oval and pinched oval-like relativistic restraint surface from Relativistic Restraint Density Theory.

The Relativistic Mass of Freely Moving Electrons and Positrons

By integrating the relativistic mass density function of a freely moving electron over the oval and pinched oval-like volumes defined by the relativistic restraint radius, $R = r_{rr}$, we can determine the enclosed mass as a function of velocity,

$$m_{rel} = \int\int\int \rho_{rel} \, R^2 \sin\theta \, dR \, d\phi \, d\theta \tag{9-15}$$

$$= \int_0^{2\pi} d\phi \int_0^\pi \int_0^{r_{rr}} \frac{K \sin\theta}{4\,\pi c \gamma \left[1 - \left(v^2/c^2\right) \sin^2\theta\right]^{3/2}} \, dR \, d\theta \tag{9-16}$$

$$= \frac{K\,r}{2\,c\,\gamma^2} \int_0^\pi \frac{\sin\theta}{\left[1 - \left(v^2/c^2\right)\sin^2\theta\right]^{9/4}} \, d\theta \tag{9-17}$$

The integral function in equation (9-17) was integrated exactly by *Mathematica* to result in the following equation for relativistic mass.

$$m_{rel} = \frac{1}{2940\,c\,v\,((c-v)\,(c+v))^{7/4}\,\text{Gamma}\!\left[-\frac{7}{4}\right]^2}\,K\,r\left(2048\,(-1)^{1/4}\,c^{5/2}\,\pi^{3/2}\,(c-v)\,(c+v) + \right.$$

$$588\,v\,((c-v)\,(c+v))^{3/4}\left(5\,c^2 - 2\,v^2\right)\text{Gamma}\!\left[-\frac{7}{4}\right]^2 - 441\,i\,c^{5/2}\,(c-v)$$

$$\left. (c+v)\left(\text{Beta}\!\left[\frac{c^2}{c^2-v^2},\frac{3}{4},\frac{1}{2}\right] - i\,\text{Beta}\!\left[1-\frac{v^2}{c^2},\frac{3}{4},\frac{1}{2}\right]\right)\text{Gamma}\!\left[-\frac{7}{4}\right]^2\right) \tag{9-18}$$

This equation from Relativistic Restraint Density Theory (RDT) is plotted with the historical L-E equation in Figure 9-3. Also included are data points from three different experiments historically used to test the L-E equation [46, p. 226-232].

In Figure 9-3, we find the resulting relativistic mass function from Relativistic Restraint Density Theory increases in value with nearly the same shape as the γ-function up to say about v/c = 0.7, and then predicts a higher mass gain at higher velocities. That is, the new theory predicts an electron has a substantially higher mass than previously thought at speeds close to the speed of light. However, this prediction is *not* confirmed by the Kaufmann data which prefers the L-E curve. I must note here that it may be inappropriate to plot together the two curves in Figure

9-3. The derivation of the mass from RDT is the electron mass as a function of velocity in a vacuum without any electric or magnetic fields propelling it in any way. Thus, the motion of the electron is in the "longitudinal" direction, i.e. in linear motion, and as discussed above, it was in fact predicted by Abraham, Lorentz, and Einstein that there is a difference between transverse and longitudinal mass due to the orientation of external forces, and the value of longitudinal mass is greater than the transverse mass of a moving electron. The three physical experiments which favor the historical equation all used either electric or electromagnetic fields to accelerate the electrons in the transverse direction causing them to radiate while the forces were applied. Since I expect these conditions change the electric field of the electron, they would cause the relativistic restraint density surface of the electron to change shape. This should result in a lower value of mass than predicted by the RDT equation for a freely moving electron.

I have not included a plot of the Lorentz-Einstein *longitudinal* mass equation in Figure 9-3 because it increases at a much greater rate than even the RDT equation, and there are no experimental longitudinal mass data to compare to them.

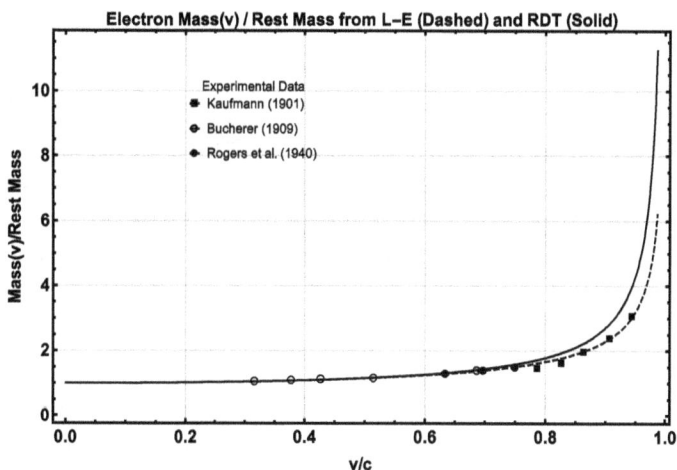

Figure 9-3. Relativistic Electron Mass(v) / Rest Mass from the Lorentz-Einstein Transverse Equation (Dashed) and Relativistic Restraint Density Theory Longitudinal Equation (Solid).

The fact that there is a difference between the two curves in Figure 9-3 brings question as to the general validity of the Lewis & Tolman equation which was derived using the conservation of momentum. I will examine the Lewis & Tolman (L&T) thought experiment in the next section.

On the Lewis and Tolman Derivation of $m_{rel} = \gamma \, m_o$

In the introduction to relativistic mechanics most textbooks discuss a dynamic thought experiment similar to that first derived by G. N. Lewis and R. C. Tolman concerning the conservation of momentum of two identical spheres in an elastic collision. In the thought experiment the two colliding spheres are assumed to have the characteristics of being solid, homogeneous, and perfectly elastic. They are sometimes referred to as billiard balls. Here are some references of textbooks containing the essence of that thought experiment: [10, p. A-13], [17, p.33], and [31, p. 64]. Assuming these conditions, the long accepted relativistic mass equation, $m_{rel} = \gamma \, m_o$, is derived.

During the derivation of this thought experiment, since the two particles are characterized as being spherical, no further consideration is given to their shape in the calculations while in motion or during the collision. The particles are assumed to remain *spherical regardless of speed*. This is surprising since by this time the textbook authors have generally already derived the Lorentz-Fitzgerald contraction showing that macroscopic spheres moving with relative velocity do not remain spherical. If this had been taken into consideration, in the motion after the collision the two spheres would not have bounced off each other at the assumed angles, and furthermore, they would have gained spin angular momentum, nullifying the derivation of the equation.

Now, in the last section I derived a new relativistic mass equation from the field-derived Relativistic Restraint Density Theory. This field theory representation of mass and its relativistic equation are expected to be exactly correct for *longitudinal* motion. Also, because of the form of the relativistic mass density function, equation (9-7), and the oval-like shapes as illustrated in Figures 9-1 and 9-2, we would have no particular reason to think *a priori* that by integrating that function over a spherical control volume, as maintained in the L&T thought experiment, that it would necessarily give us the historically accepted relativistic mass equation. However, if the assumption is made that electrons remain spherically shaped with radius r *regardless* of speed, we may expect that under these conditions the relativistic mass density function, equation (9-7), should produce the historically accepted L&T equation. And, in fact, the calculations below show that by integrating the relativistic mass density function over the at-rest *spherical* control volume of an electron, it does.

$$m_{rel} = \iiint \rho_{rel} \, dV \tag{9-19}$$

$$= \iiint \rho_{rel} \, R^2 \sin\theta \, dR \, d\phi \, d\theta \tag{9-20}$$

$$= \int_0^{2\pi} d\phi \int_0^\pi \int_0^r \frac{K \sin\theta}{4\pi c \gamma \left[1-\left(v^2/c^2\right)\sin^2\theta\right]^{3/2}} \, dR \, d\theta \tag{9-21}$$

$$= 2\pi \cdot r \cdot \frac{\gamma K}{2\pi c} \tag{9-22}$$

$$m_{rel} = \frac{\gamma K r}{c} \tag{9-23}$$

$$m_{rel} = \gamma \, m_0 \tag{9-24}$$

where m_{rel} is the relativistic mass, and m_0 is the rest mass of an electron. This derivation shows that when a point charge is *incorrectly* assumed to remain a spherically shaped object while in motion, the integration yields the Lewis & Tolman equation.

Since the equation for transverse mass of an electron developed by Lorentz under the action of a transverse force is exactly the same as the L&T equation, which was developed assuming the collision of freely moving spherical particles, I suppose it is not surprising that at that time the L&T equation was naively accepted to be correct for all motions.

Now, Einstein developed his widely accepted equation for kinetic energy *before* the L&T equation was assumed to be correct for all motions. Furthermore, Einstein derived it using his longitudinal mass equation (9-13), longitudinal mass = $\gamma^3 m_o$ [43, p. 414], not his transverse equation. So, now let's examine Einstein's kinetic energy equation, and how it compares to the kinetic energy equation derived from Relativistic Restraint Density Theory for an electron also in longitudinal motion. Fortunately, data already exists that can be used to test these two equations.

The Kinetic Energy of a Point Charge

The historically accepted equation for the kinetic energy of electrons and "ponderable masses" derived by Einstein using his longitudinal mass equation, [43, p. 414], is

$$\text{K.E.} = m_o c^2 \left(\frac{1}{\sqrt{1 - v^2/c^2}} - 1 \right) \tag{9-25}$$

Therefore, let's derive a *new* relativistic kinetic energy function for a freely moving electron, i.e. in the *longitudinal* direction, from Relativistic Restraint Density Theory for comparison. This can be accomplished using the new relativistic mass equation (9-18). As customary, to do this I will assume the kinetic energy function is equal to the work done, W, to bring a relativistic mass at rest to a speed v using the equation

$$W = \int_0^x F \, dx \tag{9-26}$$

where

$$F = \frac{d}{dt}(m_{rel} \, v) \tag{9-27}$$

Carrying out the differentiation and combining the two equations,

$$W = \int_0^x \left(m_{rel} \frac{dv}{dt} + v \frac{dm_{rel}}{dt} \right) dx \tag{9-28}$$

or equivalently,

$$W = \int_0^v \left(m_{rel} \, v + v^2 \frac{dm_{rel}}{dv} \right) dv \tag{9-29}$$

Now, we know exactly what the relativistic mass of a freely moving electron is as a function of v by equation (9-18). Taking the derivative of the relativistic mass function with respect to v, inserting it and the relativistic mass function itself into equation (9-29), and consolidating and simplifying terms, we have

$$W =$$

$$\int_0^v \frac{1}{490 \, c \, ((c-v)(c+v))^{11/4} \left(1-\frac{v^2}{c^2}\right)^{1/4} \mathrm{Gamma}\left[-\frac{7}{4}\right]^2} \, K \, r \Bigg(512 \, (-1)^{1/4} \, c^{9/2} \, \pi^{3/2} \, v^2 \left(1 - \frac{v^2}{c^2}\right)^{1/4}$$

$$-512 \, (-1)^{1/4} \, c^{5/2} \, \pi^{3/2} \, v^4 \left(1 - \frac{v^2}{c^2}\right)^{1/4} + 49 \, \mathrm{Gamma}\left[-\frac{7}{4}\right]^2$$

$$\Bigg(3 \, c^{13/2} \sqrt{\frac{v^2}{c^2}} + 3 \, c^{5/2} \, v^4 \sqrt{\frac{v^2}{c^2}} + 10 \, c^4 \, v \, ((c-v)(c+v))^{3/4} \left(1 - \frac{v^2}{c^2}\right)^{1/4}$$

$$-2 \, c^2 \, v^3 \, ((c-v)(c+v))^{3/4} \left(1 - \frac{v^2}{c^2}\right)^{1/4}$$

$$+4 \, v^5 \, ((c-v)(c+v))^{3/4} \left(1 - \frac{v^2}{c^2}\right)^{1/4} - 3 \, c^{9/2} \, v^2 \Bigg(2 \sqrt{\frac{v^2}{c^2}} + $$

$$i \left(1 - \frac{c^2}{v^2}\right)^{1/4} \left(-\frac{c^2}{v^2}\right)^{3/4} \sqrt{-\frac{v^2}{c^2}} \left(1 - \frac{v^2}{c^2}\right)^{3/4} + 3 \, i \, c^{5/2} \left(1 - \frac{c^2}{v^2}\right)^{1/4}$$

$$\left(-\frac{c^2}{v^2}\right)^{3/4} v^4 \left(1 - \frac{v^2}{c^2}\right)^{1/4} \mathrm{Hypergeometric2F1}\left[\frac{1}{2}, \frac{3}{4}, \frac{7}{4}, \frac{c^2}{c^2-v^2}\right]$$

$$-3 \sqrt{c} \, v^2 \, (c^2 - v^2)^2 \, \mathrm{Hypergeometric2F1}\left[\frac{1}{2}, \frac{3}{4}, \frac{7}{4}, 1 - \frac{v^2}{c^2}\right] \Bigg) \Bigg) dv$$

$$\tag{9-30}$$

Since the work done is equal to the kinetic energy, the new function for the kinetic energy of an electron can be plotted with the historical kinetic energy function. By solving equation (9-30) numerically, the two functions are presented together in Figure 9-4. This figure contains a graph of the historical relativistic kinetic energy from equation (9-25) (dashed line) and that derived from Relativistic Restraint Density Theory from equation (9-30) (solid line). Both equations are plotted as the square of normalized velocity versus electron kinetic energy in the typically used units of million electron volts. Also included are data listed in Table 9-1 from a frequently cited experiment using electrons to directly confirm the relativistic effect from calorimetric measurements which was conducted by William Bertozzi in 1964 [49, p. 7][50]. This data should be compatible with the RDT curve since the electrons were linearly accelerated, and then allowed to fly freely into the calorimeter. This doesn't encumber the electrons with any accelerating fields or transverse forces during the measurement. The fifth data point from that study was not plotted because the speed of light listed for that data point is impossible to be reached by electrons.

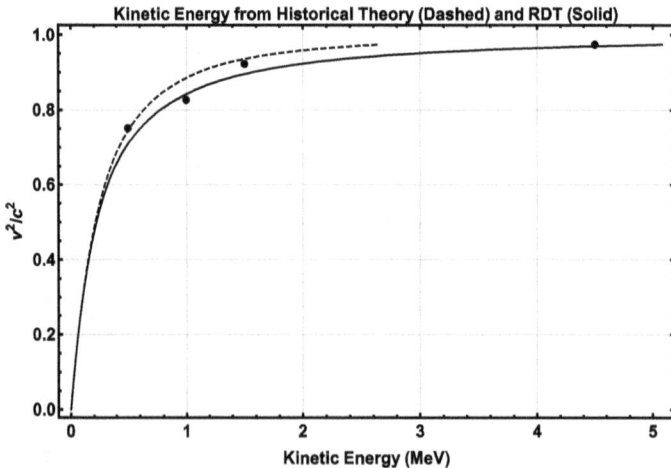

Figure 9-4. Relativistic Electron Kinetic Energy from the Historical Equation (Dashed) and from Relativistic Restraint Density Theory (Solid) with Bertozzi Data (Dots).

Here, Relativistic Restraint Density Theory predicts that an electron has more kinetic energy than predicted by the historical equation. That is, it requires more work on an electron to reach any given speed than previously thought. The two theoretical curves are each plotted from the speed of zero up to the maximum speed of $v^2/c^2 = 0.974$, where the Bertozzi data indicate the kinetic energy of 4.5 MeV was attained. Notice that the 4.5 MeV data point is much farther to the right than where it is predicted to be at the end of the dashed historical curve, and closer to the end of the solid RDT curve. The kinetic energy of this data point is much higher than predicted by the historical equation.

Bertozzi Data	K.E. (MeV)	v^2/c^2
1	0.5	0.752
2	1.	0.828
3	1.5	0.922
4	4.5	0.974
5	15	1.

Table 9-1. Bertozzi Data.

In Figure 9-5, theoretically predicted value markers, indicated by delta symbols, are located on the two theoretical curves where they each predict the Bertozzi data should be. A chart of the percentage differences between the theoretical values and the measured data are shown in Table 9-2. For example, the last theoretical point on the historical curve predicts about a 40.9 percent lower value of kinetic energy than the datum indicates was achieved, whereas RDT predicts a 12.9 percent higher value. Clearly, the RDT equation fits the data better.

Some difference between predicted and measured values may have been due to the inability to accurately measure electron speeds or kinetic energies. However, according to Edust Theory, an experiment involving calorimetric measurements such as the Bertozzi experiment, may have more difficulties than just high speed and energy measurement accuracies to determine the kinetic energy of the particles involved. Since edust is believed to be the constituent of all matter and non-matter, i.e. radiation, and we know neutrinos which are said to have no charge can easily pass through macroscopic matter without interacting with any of it, edust released by the collisions of electrons, positrons, or any other form of matter may easily pass through the surrounding material without any interaction. In such cases, those particles would not contribute any energy to calorimetric measurements. Thus, calorimetric measurements, assuming accurate laboratory equipment, should not be higher than predicted by a correct theory.

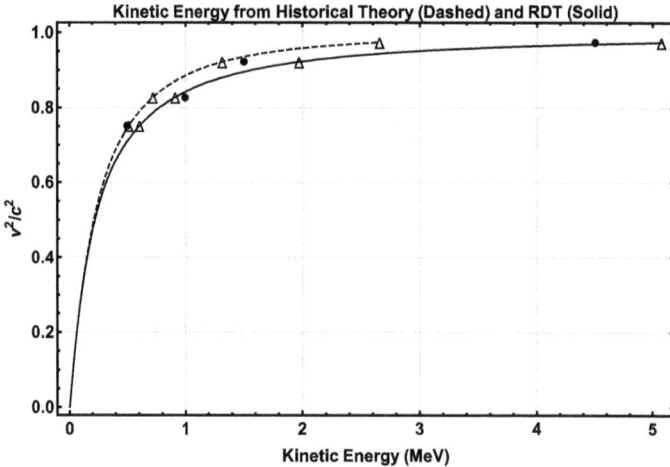

Figure 9-5. Figure 9-4 with Delta Value Markers.

Data Point	1	2	3	4
Hist. % Diff	3.03	-27.9	-12.1	-40.9
RDT % Diff	21.2	-8.82	31.6	12.9

Table 9-2. Predicted Percent Difference from Measured Value.

Because of this, Relativistic Restraint Density Theory may predict energy levels higher than calorimetric measurements. This is true for 3 of the 4 data points. For the historical energy equation, there is no significant reason why the kinetic energy should be higher than predicted, but 3 of the 4 data points exceed those predicted. Thus, the Bertozzi energy data generally appear to be understandable within RDT constraints, but not from the historical equation. That is, if small measurement errors are assumed to be allowed, all of the data are potentially explainable by RDT, but several of the data indicate much higher energy levels, particularly the -40.9 percent difference in the 4th data point, that are not explainable by the historical kinetic energy equation.

It should also be noted that the kinetic energy data listed in the table by Bertozzi were taken from theoretical values calculated from his electron accelerators, not from his direct thermal measurements. The direct thermal measurements actually indicate the higher kinetic energies of 1.6 and 4.8 MeV, rather than 1.5 and 4.5 MeV, for the third and fourth data. These would result in an even better RDT fit.

Accordingly, the conclusion in the Bertozzi publication does not make any strong claim that the data actually fits the historical theoretical equation, only of "the conformity [of] the experimental data to the Einstein relation with its prediction of a limiting speed." [50] Of course, both theories limit the velocity of macroscopic particles to the speed of light. In fact, it is easy to see now why there is this limit; all macroscopic particles are completely composed of edust particles which move at the speed of light. Therefore, no macroscopic particle can move faster than they do.

It appears that the kinetic energy equation derived here from RDT for the case of a freely moving electron is correct, and this tends to validate the RDT mass equation used to derive it. Also, the L-E equation for relativistic mass may be accurate for an electron being accelerated under the influence of a transverse force. However, RDT should be applicable to all situations. Therefore, I propose the following hypothesis to test.

Hypothesis XIII:

An edust density greater than or equal to the relativistic restraint density is the cause for matter to exhibit the property of macroscopic mass, and increase mass with speed.

Perhaps, in the future someone will be able to obtain data to directly test the RDT longitudinal relativistic mass equation, and derive an RDT equation for a relativistic electron accelerating in circular motion to determine the shape of the restraint density surface and enclosed mass as a function of speed to compare to the existing corresponding historical transversely forced data.

The past generally accepted concept of the electron has been of a particle that does not change mass with speed, but only gains kinetic energy. The phenomenon found here is a motion-dependent distribution of electron dust particles that changes density and total restrained mass with the speed of the point charge which creates it. It demonstrates that relativistic mass really exists.

According to this theory, relativistic mass may be added to phenomena such as the Lorentz-Fitzgerald contraction and time dilation of a moving clock which are not in any sense fictitious. Relativistic mass has as much strict physical significance as rest mass, length, and time duration as measured unambiguously in any inertial reference frame.

Approximations from Historical Point Particle Theories

As a consequence of the Michelson-Morley experiment many new equations were derived. Those involving solely the Lorentz kinematical coordinate transformation equations, which do not explicitly involve any mass, appear to be entirely correct. However, when macroscopic mass was associated with a point particle, i.e. rest mass, an approximation was made. As a consequence, the historical longitudinal and transverse mass equations, and all of the other equations derived from them, may be no more than approximations. This was illustrated above for the kinetic energy equation. Another example is the derivation of the rest energy of an electron. Since all of the edust particles within the rest mass of an electron are true point particles moving in rectilinear motion, it is easy to determine that the kinetic energy equation for a classical point particle applies, and the total energy within the restraint density surface of the electron at rest is $E = \frac{1}{2} m_0 c^2$, and not $m_0 c^2$ as derived with the historical relativistic L&E equation.

Absolute Space - Fixed and Perhaps Creative

As previously hypothesized and confirmed by the determination of the mass of the electron, edust enters the universe at a constant rate at the centers of sources and departs in the centers of sinks. This appears to be true for point charges while in motion, as well as at rest. It is also clear that point charges don't carry any edust or hold an unlimited quantity of edust at their centers to continuously emit those particles. Therefore, edust is either created at the centers of source point charges, or the centers activate space itself to create and eject edust. That is, the centers of positrons and electrons may behave like computer cursors which constantly activate absolute space wherever they are to cause edust to be created and destroyed, or otherwise enter and exit the universe.

Now, I had always thought of humans to be independent and autonomous, able to move about "wirelessly," carrying our energy with us with no connections to anything. However, if positrons and electrons behave like computer cursors, we are not autonomous inhabitants that move unconnected through space. Our existence may be continuously sustained from that part of absolute space in which we currently inhabit, and this would also be true for everything else.

Regardless, if all edust particles enter at the speed of light at the fixed points in absolute space wherever sources happen to be, this ensures that edust particles always move at the speed of light independent of the velocities of sources.

Also, the constancy of the speed of light, independent of the speed of its source, in all inertial reference frames was determined to be true from the Michelson-Morley experiment, but not explained. The facts that edust enters from the points in absolute space wherever sources happen to be, and our inertial reference frames are attached to matter which transforms according to the Lorentz coordinate transformation group, provide the explanation for the constant speed of light in all inertial reference frames. This also invalidates any conjectures that absolute space is warped by matter, and that gravity is caused by the warping of absolute space. Absolute space is fixed and unaffected by matter.

Chapter 10

On the Cause of Gravity

Matter Attracts Matter, but How?

Although Newton's Law of Universal Gravitation and Einstein's General Theory of Relativity describe the large scale effects of gravity, they do not explain the mechanism that causes it. However, as concluded in Chapter 4, the electron dust collision force is at the heart of all known forces. Therefore, the force of gravitational attraction between common bits of matter must be the result of the fundamental collision force of edust particles. Note that I did not say gravity attracts mass to mass, because edust particles, which are made of mass, are unaffected by gravity.

Since I proposed a definition of matter in Chapter 8, let's think about how the building blocks of matter from that definition, electrons and positrons, may exert the force of gravity on other matter. Since edust streams away from one point charge towards another or radiates to infinity, if there is a force of gravity between them it must be either caused by the edust streaming between them, or the radiation waves they cause when they are orbiting and oscillating.

Occasionally, I have read statements in physics textbooks that say two isolated point charges at rest are subjected to only the electric force "and a negligibly small gravitational force" acting between them. However, I am unaware of any direct measurements of a gravitational force between two point charges to determine a point charge's gravitational mass. Its inertial mass is usually measured in experiments with electromagnetic forces much larger than gravity acting on it. So, there may not be any evidence that a single electron or single positron is attracted at all by the force of gravity. When matter in its most elementary form, i.e. a single point charge, interacts with another point charge at large separation distances, it does so by the Coulomb force which is much larger than the predicted gravitational force roughly by a factor of 4.2×10^{42}. Therefore, it is unlikely that any gravitational force between two single point charges can ever be measured directly. Also, since individual edust particles are unaffected by gravity, and point charges are solely composed of edust, the force of gravity may not act at all between two single point charges. It may have been just a natural mistake to assume two point charges have a force of gravity between them based on the previous generally accepted notion that all particles with mass attract each other by gravity.

If this is true, let's consider another possibility, i.e. how a distant single point charge may affect matter in its next most elementary form, a positronium atom. To greatly simplify things, let's consider this first as if the two components of the positronium atom, i.e. an electron and positron, are fixed at an equal distance from a point charge, and not orbiting about their center of mass. In the analysis of Chapter 7 in Figure 7-15 the repulsive force between stationary point charges was seen to be slightly larger than the attractive force at separation distances roughly larger than 1.1 times the electron's radius. If there is no gravitational force between two point charges, the difference between the Coulomb forces of a single point charge acting on the two equally distant point charges of a positronium atom favors the larger repulsive force over the smaller attractive force. This results in a net force opposite to that of gravity. This would mean a single point charge very slightly repels a positronium atom.

Since we have not found any attractive force here, let's take the next step and consider how two fixed non-orbiting stationary positronium atoms may exert an attractive force of gravity on each other. Still, this is similar to the previous case. They may be slightly repulsed by each other. There is no reason to believe that two of those fixed atoms would be attracted to each other at large distances by any force.

Since, we know that there *is* a gravitational force acting between two pieces of common matter, and we have examined the radiation from *stationary* point charges without any evidence that they can cause gravity, let's look at the force of radiation in its more general form, i.e. oscillatory waves that propagate out to infinity.

Do Electromagnetic Radiation Waves Cause Gravity?

Since the radiation from *stationary* point charges does not show any evidence that it causes gravity, and the phenomenon of gravity can act at very long distances, I shall propose the following hypothesis:

Hypothesis XIV: *The force of gravity is a consequence of electromagnetic waves radiating from atoms.*

As discussed previously in Chapter 6, electromagnetic fields and the waves in those fields are not always composed of photons caused by electrons changing orbits, although apparently only photon wave packets are what have been found in atomic spectroscopy experiments from excited atoms. Oscillating radiation waves must also be continuously produced by unexcited atoms in stationary states due to the accelerations of their nuclei and electrons. Since unexcited as well as excited atoms are attracted by gravity, the continuous waves of radiation from atoms may be the cause of the continuous attractive effect of gravity.

Because I will generally be concerned with continuous waves of radiation by atoms in this chapter, let's consider a well documented generator of continuous radiation, an oscillating electric dipole. An electric dipole, as shown in Figure 10-1, consists of positive and negative particles of equal charge magnitude, q, oscillating about the origin on the z-axis where the motion of the charges is given by

$$z = \pm \tfrac{d}{2} \cos \omega t \qquad\qquad\qquad (10\text{-}1)$$

where ω is the oscillating angular frequency and t is time.

From these accelerating charges, the equations for the electric field, **E**, and magnetic field, **B**, of radiation at any point (R, θ) at large distances where $R \gg d$ can be approximated [23, p. 470] [51, p. 457], by the equations

$$\mathbf{E} = -\frac{qd\,\omega^2\sin\theta}{4\pi\,\epsilon_0\,c^2\,R}\cos[\omega(t - R/c)]\,\hat{\theta} \tag{10-2}$$

$$\mathbf{B} = -\frac{qd\,\omega^2\sin\theta}{4\pi\,\epsilon_0\,c^3\,R}\cos[\omega(t - R/c)]\,\hat{\phi} \tag{10-3}$$

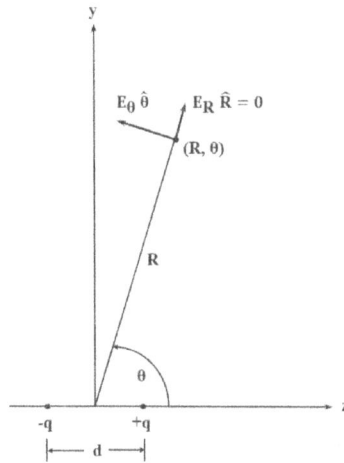

Figure 10-1. Oscillating electric dipole

The magnitudes of both the electric and magnetic radiation fields depend on a sin θ term which results in the maximum strength at $\theta = 90°$, and decrease on each side to zero at $\theta = 0°$ and $\theta = 180°$.

Now, since I have hypothesized that gravity is caused by electromagnetic radiation, the question is: How can electromagnetic radiation produce a force on matter that is roughly 4.2×10^{42} times *smaller* than the Coulomb force? To answer this question we obviously need to look for something that can produce a very small force.

Looking for Something to Produce a Small Force

In the study of electromagnetic radiation, it is normally emphasized that radiation waves in the far field are spherical, not plane waves, but for large R they are in many cases approximated as plane over small regions, similar to the surface of the earth which can be reasonably assumed to be flat locally. However, even though the surface of these waves can be approximated as flat, it is precisely the small spherical curvature at large R that I believe may cause the gravitational attraction between common bits of matter.

Note that the electric radiation field in equation (10-2) only has a component in the θ-direction. It does not have a component in the radial direction, the direction that might be thought that the field would need to act on matter to cause a gravitational effect. So, if we have the electric field of a dipole acting on our smallest bit of matter, a single point charge, the electric field will always act in the θ-direction, making it obvious that the field will simply exert a force tangent to the field, and cause the single point charge to move farther away from the center of radiation. This is an anti-gravitational effect. Once the point charge would start to move away it would have a velocity that would cause a force in the R-direction from its interaction with the magnetic field, but since the magnetic field is oscillating back and forth over time, the average effect in the R-direction would tend to be zero. There doesn't appear to be any mechanism for gravitational attraction here.

Does It Take Two?

Next, let's consider how dipole radiation may affect matter in its next most elementary form, a positronium atom. Let's also assume the radiation from the electric dipole is similar to that from a positronium atom in its first unexcited state which has no angular momentum. So, essentially we are going to examine the effect of radiation from one positronium atom acting on another positronium atom at large R as shown in Figure 10-2.

Now, we could derive the equations of motion for the four point charges in the two positronium atoms acting on each other while they orbit about their center of mass, but this leads to multiple coupled highly nonlinear, sinusoidally forced equations of motion which are very difficult to solve not only exactly, but also numerically in any conclusive or definitive manner. Since solving these equations currently seems to be prohibitive, instead let's consider heuristically a very simplified example of their interactions to propose a potential mechanism for the cause of gravity.

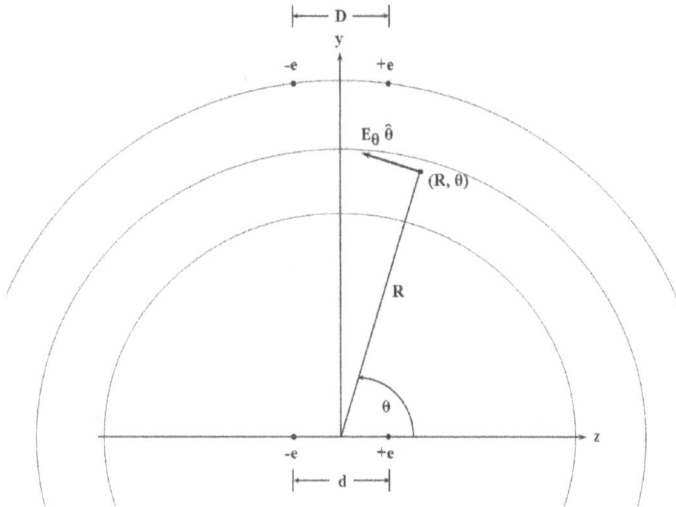

Figure 10-2. Positronium radiation impinging on a positronium atom

To begin, let's assume for the moment the positronium atom receiving the radiation does not move, but is held fixed in the current configuration. Now, let's allow the sinusoidal radiation field from the atom at the origin to propagate by it as time passes. As the radiation propagates by, it will exert a force on the positron according to the equations

$$\mathbf{F_p} = e\,\mathbf{E} \qquad\qquad (10\text{-}4)$$

$$\mathbf{F_p} = -\,\frac{e^2\,d\,\omega^2\,\sin\theta_p}{4\,\pi\,\epsilon_0\,c^2\,R}\,\cos[\omega(t - R/c)]\,\hat{\theta} \qquad\qquad (10\text{-}5)$$

and a force on the electron according to the equations

$$\mathbf{F_e} = -\,e\,\mathbf{E} \qquad\qquad (10\text{-}6)$$

$$\mathbf{F_e} = + \frac{e^2 \, d \, \omega^2 \, \sin\theta_e}{4\pi\,\epsilon_0\,c^2\,R} \cos[\omega(t - R/c)]\, \hat{\theta} \qquad\qquad (10\text{-}7)$$

In Figure 10-3, I have added another positronium atom in the far field to show the forces on their component point charges due to the sign of the electric field they are experiencing. If we assume that t and R are such that the cosine term is positive, the electric field, E_θ, acting on the more distant positronium atom is negative and causes a force on the electron to the left, and a force on the positron to the right. If the electric field is positive as acting on the nearer positronium atom, the forces would act in the opposite direction. Also, it is plain to see the curvature of the electric field causes the tangential forces on the positronium atoms to have a component of force acting at their *centers of gravity in the y-direction*. The atom in the negative electric field would be attracted toward the radiating atom, while the positronium atom in the positive electric field would be repulsed.

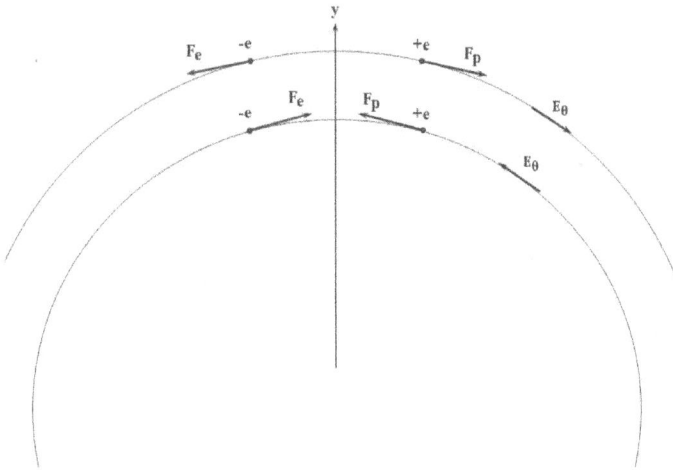

Figure 10-3. Forces due to radiation acting on two positronium atoms

Now, if the positronium atoms in Figure 10-3 were rigid, the net force on each of them in the y-direction over one cycle due to the electric field would be zero because the forces would simply alternate in opposite directions and cancel out each other. However, atoms are not rigid so the point charges are free to move during the action of the forces in the θ-direction. If the forces on the distant atom slightly pull the two point charges farther apart, the forces will rotate slightly downward increasing their angles with respect to a horizontal line, and consequently increase their force components in the negative y-direction. If the forces on the closer atom slightly push the two point charges together, those forces will also rotate slightly downward, but in this case they will decrease their angles with respect to a horizontal line, and decrease their force components in the positive y-direction. This is particularly interesting because the component forces acting in the negative y-direction become slightly *larger* when the atoms are stretched apart by the electric field, and the component forces acting in the positive y-direction become slightly *smaller* when the atoms are pushed together by the electric field. This may be the cause of a small net *quantum gravitational forc*e, *caused solely by edust collisions*.

Let's make this a little more quantitative by changing the force equations from the spherical coordinate system (R, θ) to the Cartesian coordinates (y, z), where the z-axis is pointing to the right. The unit vector in the θ-direction can be replaced by the Cartesian unit vectors by the equation

$$\hat{\theta} = \hat{x} \cos \theta \cos \phi + \hat{y} \cos \theta \sin \phi - \hat{z} \sin \theta \qquad (10\text{-}8)$$

[52, p.125], but for this case $\phi = 90°$, so

$$\hat{\theta} = \hat{y} \cos \theta - \hat{z} \sin \theta \qquad (10\text{-}9)$$

and the radiation force on the positron in Figure 10-2 becomes

$$\mathbf{F}_p = - \frac{e^2 d \, \omega^2 \sin \theta_p}{4 \pi \epsilon_0 c^2 R} \cos[\omega(t\text{-}R/c)] \, (\hat{y} \cos \theta_p - \hat{z} \sin \theta_p) \qquad (10\text{-}10)$$

Now, since we are only concerned with the forces in the y-direction that may produce a gravitational force, the y-component of the force on the positron is

$$\mathbf{F}_{py} = - \frac{e^2 d \, \omega^2 \sin \theta_p \cos \theta_p}{4 \pi \epsilon_0 c^2 R} \cos[\omega(t - R/c)] \, \hat{y} \qquad (10\text{-}11)$$

and the y-component of the force on the electron is

$$\mathbf{F}_{ey} = + \frac{e^2 d \, \omega^2 \sin \theta_e \cos \theta_e}{4 \pi \epsilon_0 c^2 R} \cos[\omega(t - R/c)] \, \hat{y} \qquad (10\text{-}12)$$

Now, from the symmetry in Figure 10-2,

$$\sin \theta_p = \sin \theta_e \qquad (10\text{-}13)$$

and since $\cos \theta = z/R$, where $z = D/2$ for the positron, and $z = - D/2$ for the electron,

$$\cos \theta_p = - \cos \theta_e \qquad (10\text{-}14)$$

By substituting the values of $\sin \theta_e$ and $\cos \theta_e$ from equations (10-13) and (10-14) respectively into equation (10-12) it can be shown that the forces in the y-direction acting on the positron and electron in the positronium atom in this configuration are identical, and the total force in the y-direction on the atom's *center of gravity* can be written as

$$\mathbf{F}_{cgy} = \mathbf{F}_{py} + \mathbf{F}_{ey} = - \frac{e^2 d \, \omega^2 \sin \theta_p \cos \theta_p}{2 \pi \epsilon_0 c^2 R} \cos[\omega(t - R/c)] \, \hat{y} \qquad (10\text{-}15)$$

Now the root mean squared value of a cosine wave is the peak value divided by $\sqrt{2}$, so the rms force acting on the center of gravity in the y-direction is

$$\text{rms } \mathbf{F}_{cgy} = \frac{e^2 d \, \omega^2 \sin \theta_p \cos \theta_p}{2 \sqrt{2} \, \pi \epsilon_0 c^2 R} \hat{y} \qquad (10\text{-}16)$$

and since $\cos \theta_p = D/2R$, and for large R, $\sin \theta_p = y/R \simeq 1$, the rms force acting on the center of gravity in the y-direction can put in the form

$$\text{rms } \mathbf{F}_{\text{cgy}} = \frac{e^2 \, d \, D \, \omega^2}{4 \sqrt{2} \ \pi \, \epsilon_0 \, c^2 \, R^2} \, \hat{\mathbf{y}} \tag{10-17}$$

Note that although the electric field that dominates the far field in equation (10-2) is proportional to $1/R$ in the θ-direction, the rms force on the center of gravity of the fixed positronium atom in equation (10-17) falls off at $1/R^2$ in the y-direction, *the same inverse R^2- dependence as in Newton's equation of gravity.*

Now, as previously stated above, if this configuration of two point charges were rigid, the net effect of the oscillating force at the center of gravity would be zero, but we know that the positronium atom is not rigid. That is, D would not be a constant, and the point charges would move slightly under the action of the forces. If we assume the point charges are displaced in phase with the forces of the electric field, then D and the magnitudes of the forces and forces' y-components would all oscillate together, becoming smaller and larger as the electric field passes by. As D becomes larger more attraction occurs and as D becomes smaller less repulsion occurs, causing a small net attractive force on the positronium atom in the negative y-direction.

This heuristic analysis of the force of radiation on a fixed positronium atom results in two of the required characteristics of the gravitational force: (1) a net average force on the center of gravity that is in the negative y-direction, i.e. attractive, and (2) it varies inversely as $1/R^2$. Obviously, however, this has been a very limited analysis with many assumptions. There is much more going on inside matter than presented in this example of a non-orbiting positronium atom. The electrons and nuclei of atoms are orbiting while the electromagnetic fields are moving past them, and there are many sizes, orientations, and phases of those orbits, as there are at the atoms from which the electric fields that excite them originate. In addition, consider ation should be given to the point charges inside of nuclei associated with the formation of protons and neutrons. They may be the predominant cause of gravity because they should have a very high orbital frequency, and would be common to all macroscopic matter which apparently has the same acceleration under the action of gravity. It seems that much needs to be discovered about the components of nucleons before this possible cause of gravity can be fully confirmed. However, for further consideration and testing the follow hypothesis is made:

Hypothesis XV:

Gravity is a consequence of atoms constantly radiating edust particle waves. Specifically, the gravitational force is caused by the spherical curvature of the electromagnetic radiation field acting on pairs of oppositely charged point charges, or their merged particles, naturally orbiting about their center of mass.

Corollary I to Hypothesis XV:

All objects producing or attracted by gravity must contain at least one positron and one electron.

Corollary II to Hypothesis XV:

Electromagnetic waves (e.g. light, radio waves, x-rays) are not attracted by gravity.

This last corollary seems reasonable since electromagnetic waves are composed solely of edust, and edust is unaffected by gravity. This also explains why black holes can radiate electromagnetic waves. However, this opposes the currently accepted theory that light is bent by gravity. I am not suggesting light is not bent, but that the cause is not gravity. I propose that light is bent by the density variations of the edust radiation fields that surround matter, just as light propagating through air bends when it passes into the higher density of water. If this line of reasoning on the nature of gravity proves to be true, it may also lead to the causes of other phenomena, i.e. Dark Energy and Dark Matter.

Dark Energy

Even though the force of gravity attracts the atoms of matter, the universe keeps expanding outward faster and faster in every direction. It is said that the far regions of the universe are moving away from us at a speed near the speed of light. To account for this, astrophysicists have proposed an invisible energy that counteracts gravity and expands the universe with an accelerating rate. They call it Dark Energy. It has been described as gravity giving up, turning negative, or something as such.

Although we first treated gravity as being caused by continuous electromagnetic waves, on a microscopic level the spraying of edust particles from a compressible source to a sink is not a continuous fluid flow. There are voids which must be present to allow for a variable density. Thus, the following cause of Dark Energy is hypothesized.

Hypothesis XVI:

"Dark Energy" occurs when the density of edust in propagating radiation waves becomes so rarefied that the waves do not act on attracting point charges associated with an atom of matter at the same time. This causes the gravitational force to cease to exist, and leaves the radiation force to act only on one point charge of an attracting point charge pair at a time.

When the edust particles only act on one point charge at a time the resultant tangential forces in the θ-direction cause each point charge to accelerate further away from the once attracting matter. This would explain why gravity ceases to act on objects at large distance, *and* the universe is expanding at an accelerating rate.

Dark Matter

Dark Matter is said to be a type of matter which has not yet been observed directly, and cannot be detected, but which interacts gravitationally with "ordinary" matter. Apparently, many galaxies would fly apart if they did not contain Dark Matter to provide additional gravitational forces to keep them together.

From the theory above, if Dark Matter exists, it consists of atoms containing positrons and electrons just like "ordinary" matter. In fact, it must be just a form of "ordinary" matter. From the earlier analysis, an electron-positron pair is the smallest known bit of matter that can create radiation which appears to cause gravity. Thus, Dark Matter may potentially consist of positronium atoms, the simplest form of matter to be able to cause a gravitational force. Under normal circumstances positronium atoms are not observable. Therefore, they or other composite particles made from them may exist in the universe, and create the gravitational effects said to be caused by "Dark Matter."

Also, if merged positronium atoms are the components of nucleons, as discussed in Chapter 8, then their radiation, which causes gravitational effects, may be part of the reason that nucleons and nuclei are created, and remain held together.

Did the "Big Bang" Ever Occur?

The "Big Bang" theory concludes that all mass was created at one point in time and space. However, we now know that mass is constantly being created at the points in absolute space where positrons (that is sources) happen to be. Also, one single explosion would not explain why the universe continues to expand at an increasing rate. However, this is explainable if gravity breaks down as discussed above when the radiation from matter becomes rarefied and causes the so-called "Dark Energy." All radiating matter could cause all observable regions of the universe beyond the point where that matter's gravity breaks down to recede, which appears to be the case. Furthermore, if absolute space were initially populated with positronium atoms, all larger forms of matter could have been formed throughout space near their points of origin. This could explain why the universe appears to be the same in all directions (isotropic), and why the measured cosmic microwave background radiation (which may be caused by high frequency radiation from the components of atoms) is roughly distributed evenly. Also, this eliminates the need for any so-called period of cosmic inflation where it is necessary for mass to move faster than the maximum speed of light.

Epilogue

I began this book with the purpose of explaining the causes of some well known phenomena, and to bring back locality, causality, and a little reality to physics by eliminating the cloud of mysticism surrounding the field of quantum mechanics. This, I also felt would confirm the unsupportability of some other theories, and provide a "start to refute it all" and "take us beyond." I believe those objectives have been accomplished. Of course, in doing so it was necessary to introduce "something more radical" which poses its own mysteries.

All of the results in this monograph evolved from a rather definitive Collision Principle, and essentially one new postulate, the existence of an uncharged particle of mass that travels at the maximum speed of light, the electron dust particle. This assumption implied that the edust particle is the only elementary particle in the universe, and all other particles are composed of edust. From the derivations which followed, the results show that the behaviors of positrons and electrons are precisely like a specifically derived compressible source and sink composed of flowing edust particles. Because of this compelling evidence, many accepted classical and modern physics theories currently published and taught today appear to be either only approximations or completely unsupportable.

Many other results and conclusions were derived from this evidence, some of which are listed in the Introduction. I hope that I have justified them to your satisfaction, and you feel it was well worth your time to read the book. The theories introduced appear to have the potential to help unlock more of the secrets of nucleons and the universe. Perhaps, some day specific stable configurations of orbiting electrons and positrons or their merged particles will be found which are, in fact, protons and neutrons. Perhaps, even a way to manipulate point charges will be found to easily fulfill the dreams of the medieval alchemists. In any event, hopefully whatever practical technologies using this knowledge are developed, they will be used for peaceful purposes to benefit all mankind.

Selected References

[1] Wolfram Research, Inc., *Mathematica* Versions 3-11, Champaign, Illinois.

[2] Ford, K., *The Quantum World* (Harvard University Press, Cambridge, Massachusetts, 2004)

[3] Einstein, A., *Relativity: The Special and General Theory* (1920), Translated: Robert W. Lawson (Authorised translation)

[4] Feynman, R. P., *QED: The Strange Theory of Light and Matter* (Princeton University Press, Princeton, New Jersey, 2006)

[5] Smolin, L., *The Trouble with Physics* (Houghton Mifflin Company, New York, 2006)

[6] 't Hooft, G., *In search of the ultimate building blocks* (University Press, Cambridge, United Kingdom, 1998)

[7] Woit, P., *Not Even Wrong* (Basic Books, New York, 2006)

[8] Unzicker, A., *The Higgs Fake* (Lexington, KY, 2013)

[9] Unzicker, A., and Jones, S., *Bankrupting Physics* (Palgrave Macmillan, New York, 2013)

[10] Eisberg, R., and Resnick, R., *Quantum Physics of Atoms, Molecules, Solids, Nuclei, and Particles* (John Wiley & Sons, New York, Second Edition, 1985)

[11] Merzbacher, E., *Quantum Mechanics* (John Wiley & Sons, Third Edition, 2004)

[12] Kumar, M., *Quantum* (W.W. Norton & Company, New York, 2010)

[13] Aczel, A., *Entanglement* (John Wiley & Sons, Ltd., England, 2003)

[14] Clegg, B., *The God Effect* (St. Martin's Press, New York, 2006)

[15] Griffiths, D. J., *Introduction to Quantum Mechanics* (Pearson Education, Inc., Second Edition, Upper Saddle River, New Jersey, 2005)

[16] Zee, A., *Quantum Field Theory in a Nutshell* (Princeton University Press, New Jersey, Second Edition, 2010)

[17] Beiser, A., *Perspectives of Modern Physics* (McGraw-Hill, Inc., New York, 1969)

[18] Jammer, M., *Concepts of Force* (Dover Publications, Inc., 1999)

[19] Klauber, R. D., *Student Friendly Quantum Field Theory* (Sandtrove Press, Fairfield, Iowa, Second Edition, 2013)

[20] Falkoff, D. L., *Exchange Forces* (American Journal of Physics 18 (1): 30–38, 1950)

[21] Cajori, F., *Newton's Principia* (University of California Press, Fifth Printing, 1975)

[22] Oerter, R., *The Theory of Almost Everything* (Pi Press, New York, 2006)

[23] Griffiths, D. J., *Introduction to Electrodynamics* (Pearson Education, Inc., Fourth Edition, Boston, 2013)

[24] Sakurai, J., and Napolitano, J., *Modern Quantum Mechanics* (Addison-Wesley, San Francisco, California, Second Edition, 2011)

[25] Pao, R. H. F., *Fluid Dynamics* (Charles E. Merrill Books, Inc., Columbus, Ohio, 1967)

[26] Committee on Data for Science and Technology, 2010.

[27] Veltman, M. J. G., *Facts and Mysteries in Elementary Particle Physics* (World Scientific Publishing Co. Pte. Ltd., Singapore, 2003)

[28] Schey, H. M., *div grad curl and all that* (W. W. Norton & Company, New York, Fourth Edition, 2005)

[29] Wolchover, N. and Emspak, J., *The 18 Biggest Unsolved Mysteries in Physics*, http://www.livescience.com/34052-unsolved-mysteries-physics.html, 2019.

[30] Feynman, R. P., *The Character of Physical Law* 129 (BBC/Penguin, 1965)

[31] Weidner, R. and Sells, R., *Elementary Modern Physics* (Allyn and Bacon, Inc., Boston, 12th printing, 1967)

[32] Dirac, P. A. M., *The Principles of Quantum Mechanics* (Oxford University Press, Fourth Edition, New York, 1982)

[33] Pais, A., *Inward Bound: Of Matter and Forces in the Physical World* (Oxford University Press, New York, 1988)

[34] de Broglie, L., *The wave nature of the electron* (Nobel Lecture, 1929)

[35] Weidner, R., and Sells, R., *Elementary Classical Physics Volume 2* (Allyn and Bacon, Inc., Boston, 8th printing, 1971)

[36] Close, F., *Antimatter* (Oxford University Press, Inc., New York, 2009)

[37] Close, F., *Neutrino* (Oxford University Press, Inc., New York, 2010)

[38] http://hyperphysics.phy-astr.gsu.edu/hbase/particles/qevid.html#c1, 2019.

[39] Laughlin, R. B., *Phys. Rev. Lett.* 50 (1983) 1395

[40] Störmer, H. L., *Nobel Lecture: The fractional quantum Hall effect* (Copyright © The Nobel Foundation 1998)

[41] Schwartz, H., *Introduction to Special Relativity* (McGraw-Hill Book Company, New York, 1968)

[42] Rindler, W., *Introduction to Special Relativity*, Second Edition (Oxford University Press, Inc., New York, 2003)

[43] Miller, A. I., *Albert Einstein's Special Theory of Relativity* (Addison-Wesley Publishing Company, Inc., Reading, Massachusetts, 1981)

[44] Okun, L.B., *The Concept of Mass* (PDF) (Physics Today, **42** (6): 31-36, June 1989)

[45] Lorentz, H. A., Slightly revised English translation: *Simplified Theory of Electrical and Optical Phenomena in Moving Bodies, Koninklijke Akademie van Wetenschappen te Amsterdam. Section of Sciences. Proceedings* **1** (1898-99): 427-442.

[46] Zhang, Y. Z., *Special Relativity and Its Experimental Foundations*, (World Scientific Publishing Co. Pte. Ltd., Singapore, 1997)

[47] Lewis, G. N. and Tolman, R. C., *The principle of relativity and non-Newtonian mechanics, Phil. Mag.* **18**, 510-523 (LVII 1909)

[48] Tolman, R. C., *The Theory of the Relativity of Motion* (University of California Press, Berkeley, 1917)

[49] French, A. P., *Special Relativity* (M.I.T. Introductory Physics Series, Massachusetts, 1968)

[50] Bertozzi, W., *Am. J. Phys.*, **32**, 551-555 (1964)

[51] Hazen, W. E. and Pidd, R. W., *Physics* (Addison-Wesley Publishing Company, Inc., 1965)

[52] Arfken, G. B. and Weber, H. J., *Mathematical Methods for Physicists* (Elsevier Academic Press, Sixth Edition, 2005)

Index

About the Author

Donald W. Caldwell is a retired mechanical engineer. He spent most of his career conducting research and development, and managing such projects for the United States Department of Defense. He earned his B.S., M.S., and Ph.D. degrees from the University of Maryland Department of Mechanical Engineering, while he was a teaching assistant and Minta Martin Fellow.